THE
DOOMSDAY
BOOK

MARSHALL BRAIN
CREATOR OF HOWSTUFFWORKS.COM

STERLING
New York

STERLING
New York

An Imprint of Sterling Publishing Co., Inc.
122 Fifth Avenue
New York, NY 10011

ISBN 978-1-4549-3996-2

Distributed in Canada by Sterling Publishing Co., Inc.
c/o Canadian Manda Group, 664 Annette Street
Toronto, Ontario, Canada M6S 2C8
Distributed in the United Kingdom by GMC Distribution Services
Castle Place, 166 High Street, Lewes, East Sussex, England BN7 1XU
Distributed in Australia by NewSouth Books
University of New South Wales, Sydney, NSW 2052, Australia

For information about custom editions, special sales,
and premium and corporate purchases, please contact
Sterling Special Sales at 800-805-5489
or specialsales@sterlingpublishing.com.

Manufactured in Spain

2 4 6 8 10 9 7 5 3 1

sterlingpublishing.com

Interior design by Christine Heun
Cover design by Elizabeth Mihaltse Lindy
A complete list of illustration credits appears on page 270.

This book is lovingly dedicated to my very patient wife, Leigh, and to our four children, David, Ian, Irena, and Johnny, who all played a part in this book's creation. Thank you so much for everything you do to make my life so wonderful.

CONTENTS

INTRODUCTION

Imagine that you can get into a time machine and revert back in time to 2001. You are now living in New York City, and the date is September 11. This will become a terrifying, horrific, mind-boggling day for you. But it is 8:00 in the morning, so you don't know that yet.

September 11, 2001, is a Tuesday. Chances are that you are either starting your day at work or are navigating the morning rush hour en route to your place of employment. Along much of the East Coast, the morning is beautiful and sunny. Though the leaves have not yet started to change, fall is definitely in the air.

I can distinctly remember this day. I was in Raleigh, North Carolina, at the headquarters of HowStuffWorks.com. Around 9:00 a.m., one of the employees who had been listening to the radio said, "Something is happening in New York."

Today, everyone expects the internet to let them see what is happening anywhere in the world within seconds. But this is 2001. Facebook will not launch until 2004. YouTube will not exist until 2005. Twitter will not appear until 2006. The iPhone 1 and its easy-to-use camera will not arrive until 2007. The early 2000s were a different era in the media world compared to today.

In 2001, if you really needed to know what was going on in something close to real time, you turned on a television. So we gathered at the TV we had in the office and started watching. And even in 2001, the coverage is instantaneous, because the news is breaking in New York City—the media capital of the world.

We did nothing else that day. So many people in the United States and around the world spent that Tuesday near a television. We sat there in indescribable shock, watching the horrifying, terrifying reality of the attacks unfold for everyone on live TV.

If we look at a timeline of events as they unfolded on 9/11, we can see why the attention of the entire United States started turning toward New York City around 9:00 Eastern on Tuesday morning. The first hijacked airplane hit the 104-story World Trade Center North Tower at 8:46 a.m. A gigantic, unmistakable black plume of smoke starts pouring into the crystal blue sky for all to see.

And really, we have to give the terrorists credit here. They have picked the perfect day weather-wise to destroy two of the world's most iconic buildings on live television. They have chosen a perfect time on that day for the attack to start. And they have chosen the best city in the world to garner media attention. Once this first airplane hits the North Tower and the smoke starts billowing out, New York's entire international media apparatus swings into action with laser-beam focus. The apparatus starts filming and broadcasting and commenting within minutes.

While the first airplane came out of nowhere and only one camera on the entire planet had captured it as it crashed into the North Tower, a whole flock of cameras are watching the second hijacked airplane hit the WTC South Tower at 9:03 a.m. This second attack is captured in real time. Everyone who has tuned in can see it happen live, or they can watch the infinite number of replays that will occur all throughout the day.

This second explosion is impossible to imagine . . . impossible to believe. An entire passenger jet with 65 people onboard has been recommissioned for use as a gigantic jet-fuel bomb. This second plane has perhaps 50,000 pounds of jet fuel in its tanks when it slams into the South Tower. And now two things have happened: both towers are now billowing smoke and flames into the perfect sky, and everyone suddenly understands exactly what had happened to the North Tower a few minutes earlier. Prior to 9:03 a.m., no one really knew for sure if it had been an accident or not. Now everyone knows that the nation is under attack.

At 9:42 a.m., the FAA grounds every flight in the entire United States. There are no more takeoffs, and all planes in the air have to land immediately wherever they happen to be. This is unprecedented and unbelievable—it has never happened before or since.

Then the unthinkable happens. At 9:59 a.m., with most people on the West Coast now awake and getting ready for work themselves, the South Tower collapses in the most spectacular, incredible way possible.

At 10:07 a.m., United Airlines Flight 93 nose-dives into the ground in Pennsylvania. This flight was headed for the US Capitol Building. The North Tower collapses at 10:28 a.m. A plane hits the Pentagon at 10:50 a.m., but this is revealed more slowly during the day because cameras were not on hand to film the attack live. By 11 a.m., New York City starts evacuating over a million people from Lower Manhattan. The city has to do it—who knows what else might be coming?

The avalanche of events on this day is nearly impossible to process. The average person has never contemplated or imagined what is happening, and things are unfolding so fast that it is overwhelming. And this is not an imaginary story from a superhero movie. This is real life.

At 5:20 p.m., while it is still sunny and bright for

ABOVE Smoke billows from the North and South Towers of the World Trade Center after they are struck by passenger planes on the morning of September 11, 2001.

all the cameras to see, the forty-seven-story building at World Trade Center 7 completely collapses—a final, brutal climax for the day. This is the apocalypse come to life. This is hell on Earth. This is doomsday.

• • •

The events on 9/11 are arguably the most impactful and destructive one-day conventional attacks in world history.

You might think that Pearl Harbor deserves the crown. But only 2,403 died in the Pearl Harbor attack, compared to 2,996 on 9/11. Only six warships were sunk in the Pearl Harbor attack, while three gigantic, iconic, and incredibly valuable civilian buildings (WTC1, WTC 2 and WTC7) collapsed on 9/11, right in the heart of one of America's most important cities, and the Pentagon lost one of the pents in its outer rings. And while the Pearl Harbor attack was carried out by thousands of Japanese members of their armed forces, including a large part of the Japanese Navy, and 353 Japanese airplanes, backed by the entire military might of the country, 9/11 was carried out by nineteen people armed with box cutters, on a budget of several thousand dollars.

How could this possibly happen? Prior to 9/11, the average American might have thought, probably with complete confidence, that the world's greatest military power in all of history could protect its own headquarters against any attack. But more than four hundred billion dollars in annual military spending still left the nation's preeminent city, as well as the Pentagon, in the nation's capital, completely vulnerable and defenseless.

Let's set aside the horror of the 9/11 attacks for just a moment and allow the unemotional, rational part of our brains to analyze this attack. When viewed in this light, the attacks were incredibly efficient. With a tiny force of just nineteen people and a tiny investment in their training, a small terrorist organization working out of Afghanistan and Pakistan successfully executed an unprecedented and unbelievably powerful attack on one of the world's dominant superpowers.

How is such a scenario even possible? Quite simply, this tiny organization found a chink in the superpower's armor and then exploited this chink to

the maximum extent possible. They also crafted the attack for maximum psychological impact, choosing a site where much of the attack was broadcast live to the world by the planet's best-equipped media organizations.

There were many immediate effects of 9/11. Thousands of people died, and four commercial airliners and all of their passengers and crew were lost. Three gigantic buildings worth billions of dollars were completely leveled, and the worldwide headquarters of the United States military was successfully attacked. One of the busiest transit hubs in the world, containing the World Trade Center subway and PATH stations, was destroyed and would not reopen at full capacity for seventeen years. There was an enormous stock market crash and a ripple effect across the entire US economy.

In the multi-year aftermath of 9/11, we saw many more repercussions, which we can now understand with the benefit of 20/20 hindsight. All kinds of changes and safety measures have been put into place, with their own significant costs in terms of time and money. The best-known aftereffect is the creation of the TSA (Transportation Safety Administration), which travelers now see and deal with at every airport. Passengers can no longer carry most knives, especially box cutters, onto airplanes. Cockpit doors are now armored and locked, with protocols meant to prevent hijackings. There is a no-fly list maintained by the Terrorist Screening Center that prevents tens of thousands of people from getting on a plane. The Federal Air Marshal Service (FAMS) ramped up quickly, hiring thousands of new air marshals. FAMS alone costs approximately a billion dollars per year to operate. The TSA runs about $8 billion per year.

And don't forget the wars in the Middle East. The United States subsequently invaded Afghanistan and

Iraq, with unfathomable costs in lives (approximately 500,000 lives lost) and money (upwards of $2 trillion), with no real benefit to speak of. These wars would eventually open the door for ISIS to emerge in Syria and Iraq, displacing millions of people, particularly from Syria, and causing an enormous refugee crisis in Europe, Turkey, and Lebanon, among other countries.

All this terror, war, mayhem, money, destruction, and death was unleashed by just nineteen people.

Is 9/11 a doomsday scenario? Yes, it is, absolutely. In this book, we define a doomsday scenario to be any event with the potential to cause widespread catastrophe in the form of death, destruction, and/or economic effects. In the case of 9/11, the attacks initially affected primarily New York City, but they later had far-reaching effects in many other ways. Similarly, a big tsunami can destroy billions of dollars in buildings and infrastructure, kill hundreds of thousands of people, and have severe economic consequences when important factories and power plants are destroyed.

Often these scenarios unfold because we as humans have made an assumption, and events conspire to violate the assumption in a fundamental way. During 9/11, we assumed that bad actors would not kill the pilots and recommission passenger jets as flying bombs. Now we have learned otherwise and use armored cockpit doors to keep bad actors out. In the case of earthquakes, we assume that the earth is stable. But earthquakes violate this assumption occasionally, so we build earthquake-proof buildings. Before a pandemic like the one that spread COVID-19 (see page 73), we assumed it was good to gather groups of humans together in restaurants, gyms, stadiums, and movie theaters. The virus violated this assumption and caused an economic catastrophe when millions of businesses had to temporarily close. If we

can understand these assumptions ahead of time and then mitigate their ramifications, we can often avoid disaster.

In this book, we are going to look at twenty-five potential doomsday scenarios that upend our assumptions, often in destructive ways. Some scenarios affect just a city, as is frequently the case for an earthquake (see page 142). Other scenarios can affect our entire planet: for example, an asteroid strike (see page 122). We'll also cover everything in between. Each chapter has a Threat Level indication that identifies the scope of the scenario (city, country, continent, or world).

We will look at man-made disasters (those that are initiated by humans, often with the intention of destroying things), natural disasters (those that come from natural causes, like hurricanes or volcanoes, or that disrupt nature, as is the case of rainforest collapse or ocean acidification), and several situations imagined by science fiction. For each scenario, we will look at what might cause it and what could happen as a result. Many chapters open with a dramatization that asks you to imagine yourself in the doomsday scenario or use a fictional portrayal to bring the scenario to life.

Most importantly, we will also try to examine strategies, when possible, that might be used to prevent these scenarios from ever happening. If we, as an intelligent species, take the time to imagine and study doomsday scenarios before they actually happen, there are many active steps we can take that would prevent them from ever unfolding in the real world. Think of the described scenarios in this book like a vaccination, where pre-exposure to a problem can help our society build immunity, or at least resilience.

Let's get started . . .

PART I
MAN-MADE DISASTERS

SPLITTING THE UNITED STATES IN HALF

THREAT LEVEL: COUNTRY

The events from this morning's news stunned everyone in the United States—and probably half the people on the planet. As best anyone could tell, the dam at Fort Peck Lake had spontaneously collapsed at about 9:00 last night. The details about why it had collapsed were unclear—no one really knew what had happened or why this massive earthen dam would have suddenly given way on a random April evening.

Most people who heard this news started with the same basic question: "What is Fort Peck Lake?" Fort Peck Lake is a short drive south of a small farming community in Montana, way off the beaten path. The average American has never heard of the place.

But it turns out that this lake is important. Fort Peck Lake had been the fifth largest man-made lake in the United States. The earthen dam for the lake was holding back 6 trillion gallons (22 trillion l) of water from the Missouri River. How much water is that? Here's one perspective: If all 330 million people in the United States drank a gallon (3.8 l) of water per day from this lake, it would take fifty years to drink the entire amount. One arm of the lake had stretched 20 miles (32 km) to the south. The other had stretched 40 miles (64 km) toward the west. When you stood on the shore of the lake, you could not even see the other end, because it was hidden by the curvature of the Earth. You get the idea: this lake was *big*.

According to the news reports and the helicopter footage from this morning, the dam had almost completely disappeared when it collapsed, and the lake was gone too. But that was only a tiny part of the story. The sensational, unbelievable part was what the experts were predicting for the coming days. This one dam would start a chain reaction, and that chain reaction would eventually cut the United States in half, destroying a whole series of large American cities in the process. It may take three or four days for the entire reaction to unfold, but there is not a thing that anyone can do to stop it.

PREVIOUS SPREAD The Hiroshima Prefectural Industrial Promotional Hall was the only structure that remained standing in the area where an atomic bomb was detonated on August 6, 1945. The area leveled by the bomb was about 2 miles (3.2 km) in diameter. RIGHT The dam holding back Ft. Peck Lake is essentially an enormous, engineered pile of dirt. The base of this pile is about 1 mile (1.6 km) wide and 4 miles (6.4 km) long.

The doomsday scenario that surrounds Fort Peck Lake is both surprising and fascinating. To understand it, we have to direct our attention to the linchpin in this whole equation, the thing that makes Fort Peck Lake the starting point of a gigantic catastrophe: the lake's earthen dam, built in 1937 by the US Army Corps of Engineers.

An earthen dam is just what we would expect—it is a big pile of dirt. It is an engineered pile, but a pile nonetheless. There is no question that the dam at Fort Peck is big, nearly a mile (1.6 km) thick at the base and almost 4 miles (6.4 km) long. The core of the dam is made of clay that is impermeable to water, with sand and gravel banked against the core on either side for support. The steel wall at the bottom of the dam is called a cut-off. It prevents water from leaking below the dam and weakening it.

Fort Peck Lake is a recreational site. In other words, this lake is not protected in any significant way. It is out in the open, available for everyone to use. A publicly accessible road runs along the top of the dam. It also contains a hydroelectric power plant that produces nearly 5 million kilowatt-hours of electricity per day

This means that the dam is vulnerable to attack from several different angles. A person could attack the dam by truck, by boat, or underwater. Charges planted along the dam on the lake side might be able to destroy it. An unexpected earthquake of sufficient intensity might do the trick as well. A suitcase nuclear device (see page 27) in the trunk of a car could cause the necessary damage. Whatever the mechanism of destruction, the outcome of this dam's collapse is the same. Once the dam is breached, the chain reaction that could unfold is truly epic.

The 6 trillion gallons of water in Fort Peck Lake is an immense amount by any measure. Once this mass of water finds an opening in the dam to flow through,

BELOW The cross section for the earthen dam at Fort Peck Lake shows a clay core that it impermeable to water. It is buttressed by huge piles of sand on either side. The steel piling below the dam prevents water from undermining it.

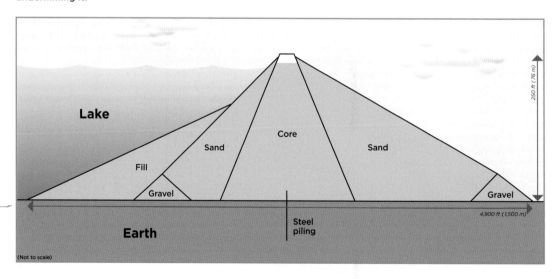

DAM FAILURES

Dam failures happen more frequently than we might expect. Here are three examples:

St. Francis Dam: In 1928, a 185-foot (56 m) concrete dam called the St. Francis Dam near Los Angeles failed and unleashed 12 billion gallons (45 billion l) of water toward downstream communities. Hundreds of people died as the wall of water swept through the area.

Teton Dam: Constructed in Idaho in 1972, this earthen dam failed just four years later as water seeped into the dam itself and undermined it, causing a catastrophic failure and the release of 90 billion gallons (340 billion l). The cost of the total damage may have been as high as $2 billion.

Oroville Dam: In 2017, massive erosion occurred in the main spillway and the emergency spillway, threatening the integrity of the dam. Nearly 200,000 people were evacuated in northern California in case the dam failed. Fortunately, the rain that caused the erosion stopped before the dam could collapse, so repairs could be made.

BELOW Water spills over a damaged Orville Dam spillway.

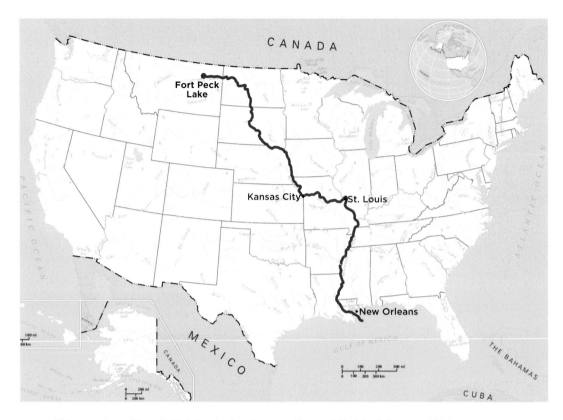

ABOVE This map shows the path that floodwaters from a collapse of Fort Peck Dam could take.

it would quickly carve out a wide passage, creating a flood of water that begins heading downstream as quickly as gravity can carry it. The "stream" in this case is the Missouri River. It is along this river that a multiplying effect takes hold and fuels a chain reaction.

Several hours after the Fort Peck Dam collapses, the flood would reach Lake Sakakawea, near Riverdale, North Dakota. Lake Sakakawea, the fourth-largest man-made lake in the United States, is slightly larger than the Fort Peck Lake. The incoming flood would likely overwhelm its dam, causing it to collapse as well. Now the amount of water in the flood has doubled.

The flood would quickly reach Bismarck, the capital of North Dakota, just 55 miles (88 km) away. The now-12 trillion gallons of water would bury Bismarck and wipe the city to its foundations.

After a few more hours, the flood would reach Lake Oahe near Pierre, South Dakota. This lake, the third largest man-made lake in the United States, is larger than Fort Peck Lake, with a capacity of 7.6 trillion gallons (29 trillion l). The immense amount of incoming water would likely overwhelm this third dam as well.

The 20 trillion gallons (76 trillion l) of water heading down the Missouri River channel would take out every small town in its path. This amazing pulse of water would destroy three smaller hydroelectric dams at Lake Sharpe, Lake Francis Case (which has the tenth largest dam of this type in the United States), and Lewis and Clark Lake. The contents of each lake cause the flood to grow even larger as it rushes past. Sioux City, Iowa, and Omaha, Nebraska, would vanish. Several hours later,

ABOVE In 2019, the Mississippi River flooded Pacific Junction, Iowa—an example of the destruction that seasonal flooding of the river can cause.

this unimaginably gigantic flood of water reaches the Kansas City metropolitan area where the Missouri River flows right through the center of town.

The metropolitan area of Kansas City has a population of about 2 million people. They would have approximately a day after the Fort Peck Dam is breached to evacuate the city, assuming that the evacuation starts the instant the first dam collapses. It would take the flood less than half an hour to scrub Kansas City from the face of the Earth. The passing deluge would pick up a huge amount of additional debris in the process.

The Missouri River merges into the Mississippi River at St. Louis, home to about 3 million people. Fort Peck Lake is approximately the size of St. Louis, and it reaches a depth of about 250 feet (76 m). The amount of water has now more than tripled since the first dam was breached in Montana. Like Kansas City, St. Louis easily disappears under this enormous wave of water. All this water and debris then enters the Mississippi River basin and starts moving toward the south.

If the Fort Peck dam happens to collapse in the springtime, the Mississippi River will be flooding already. This 20+ trillion gallons (76 trillion l) of water, plus the gigantic pile of debris created by the elimination of several major cities and countless smaller establishments, will flow down the Mississippi River toward the Gulf of Mexico, destroying Memphis, Baton Rouge, and New Orleans, along with many small towns and settlements, in the process. The water will also spread out into the Mississippi River's historical floodplains, inundating thousands of square miles of farmland and covering it in debris—millions of crumpled houses, apartments, shopping centers, uprooted trees, cars, appliances, and much more.

The loss of five major US cities and innumerable smaller ones would be significant. The populations of the Omaha, St. Louis, Kansas City, Memphis, and New Orleans metro areas alone total roughly 8 million people. Hundreds of thousands of people could die at a minimum, but the death toll would depend on our ability to evacuate large urban centers and smaller towns on short notice. If the evacuations are bungled, there is a potential for millions of deaths. Millions of survivors would also be left homeless and destitute in an instant.

As devastating as these losses would be, an equally troubling effect would descend on the United States' infrastructure. In essence, the United States would be cut in half, creating a "United States East" and a "United States West." Every bridge bringing every road, highway, and railroad across the Missouri and lower Mississippi Rivers would be erased. Power lines, pipelines, and internet lines crossing the rivers would also be lost. The United States would become two separate regions until repairs could be made, and it is easy to imagine that several years could elapse before the damage is fixed and all of the debris is cleared. Think about it this way: many imported goods from China arrive on the West Coast of the United States and are transported to the East via truck and train. A great deal of food, and many other products, follows the same routes in both directions. If all the highway and railroad bridges in the middle of the United States are erased, it creates a significant logistical challenge for the country.

SCIENCE

Why does the water in a lake have so much destructive power? It has to do with the potential energy stored in the water. A gallon (3.6 l) of water at Fort Peck Lake is approximately 2,200 feet (670 m) above sea level and weighs about 8 pounds (0.6 kg). If we drop a gallon of water through this 2,200-foot distance, we release approximately 26,300 joules, equivalent to 7 watt-hours of electricity. It is roughly the amount of energy held in a modern smartphone battery.

This may not seem like much energy, but the failure of the dam at Fort Peck starts a chain reaction, a cascade of dam failures, eventually unleashing more than 20 trillion gallons of water over the course of a few hours. While one gallon of water is pretty harmless, 20 trillion gallons creates an enormous multiplier effect. This amount of rushing water represents the equivalent of 5.5×10^{17} joules, or 550,000,000 gigajoules of energy, released once it all reaches sea level at the Gulf of Mexico. It is the water's fall from 2,200 feet in elevation to zero feet in elevation that releases the potential energy and unleashes the destructive force.

To put this into perspective, one ton of TNT holds approximately 4 gigajoules of energy. This means that the flood from the breached dams has the energy equivalent of 137,000,000 tons of TNT. The atomic bomb dropped on Hiroshima (see Nuclear Bombs, page 22) measured approximately 63,000 gigajoules, or 15,000 tons of TNT, give or take, and it easily leveled an entire city. Therefore, the potential energy stored by the lakes along the Missouri River is equivalent to thousands of Hiroshima bombs. This amount of pent-up energy, released in less than a day, would have a devastating effect on everything it encounters. It is not an intense and instantaneous burst like a nuclear bomb blast, but it is still capable of doing an enormous amount of damage.

The cascading effect is also important to keep in mind. It is something that engineers work to avoid at all cost. A cascade of failures occurs when a failure at one point in a system overstresses other

ABOVE The spillway at Fort Peck Dam is lined with concrete and helps divert excess water from Fort Peck Lake.

points in the design, leading to additional failures. Cascading failures happen in many different ways. If a power grid is near capacity and one big transmission line fails, it can cause failures in other transmission lines as they become overloaded because of the first failure (see EMP Attack, page 44). If one section of a bridge fails, the entirety of the bridge may fail in a cascade if it is not designed properly. When 7 World Trade Center collapsed during the 9/11 attacks, some theorized that a single column failure caused many subsequent failures and destroyed the entire building. Shyam Sunder, the National Institute of Standards and Technology's lead investigator of the incident, called it "a new kind of progressive collapse."

The typical strategy to avoid cascading failures is to overbuild, or to create excess capacity; this way, a failure of one component can easily be withstood by the remaining components. For example, if a power grid has twice the capacity it needs, even at peak loading, then the failure of one transmission line out of a dozen has no effect whatsoever. The remaining transmission lines would have more than enough capacity to absorb the failure. The downside is that this extra capacity costs money. But in many cases, the cost of overcapacity is far lower than the cost of a failure.

You might be wondering what we can do, given that Fort Peck Lake contains the energy equivalent of thousands of Hiroshima bombs. If this one lake can start a cascade of failures capable of destroying five major cities, killing or displacing millions of people, and cutting the country in half, how do we prevent catastrophe?

One possibility: we can look at our treatment of nuclear power plants for guidance. If a nuclear power plant (or the on-site stores of spent nuclear fuel) were somehow breached, and if its nuclear material were somehow released into the environment, we would see a doomsday scenario similar to the Chernobyl disaster in 1986. Therefore, we take extraordinary steps to protect our nuclear power plants.

For example: In 1998, anyone could visit Raleigh's Sherron-Harris nuclear power plant. Sherron-Harris is where the electricity for Raleigh, North Carolina, primarily comes from. In that era, pre-9/11, any civilian could drive right into the parking lot next to the nuclear reactor and start snapping photos. There was not much concern at all about people sabotaging or attacking nuclear power plants.

Today, we take nuclear power plant security very seriously. As a result of events like September 11, 2001, there are armed guards, fences, gates, background checks, and more at every nuclear power plant. Civilians can no longer come anywhere near a nuclear reactor site or a nuclear waste site.

Therefore, if Fort Peck Lake contains the energy-equivalent of a thousand Hiroshima bombs, it seems logical to implement similar measures:

- Ban cars, trucks, boats, and airplanes from going anywhere near the dam. If airplanes happened to break into the restricted airspace, they would be shot down.

- Build sonar systems in the lake to detect scuba divers, small submarines, and any other underwater activity.

- Consider safeguards against people or robots digging hidden tunnels into or underneath the dam from either side.

- Reduce the amount of water in the lake.

- Build a second dam downstream from Fort Peck Dam to catch the water if the first dam fails. This second dam would add redundancy to the system of dams. Similar secondary dams can be constructed at many of the man-made lakes in the United States, especially those with earthen dams.

- Devote sufficient resources to understand the dam's weaknesses and maintain the dam, especially as the dam ages and the need for repairs grows.

The steps listed above all seem like logical, commonsense things to do, given the enormity of the threat that the dam represents. And to be fair, some of these things are being done. But you can look at aerial photos of the dam right now and see that there are cars and trucks driving over the dam. You can use Google Street View to see that the lake's water level is right up next to the road. You can look at the Federal Aviation Administration's charts and see that there are no restrictions on flying near the lake. These all create unnecessary vulnerabilities, especially for a dam that is holding back this much potential energy.

Mass destruction often happens when sites that store large quantities of potential energy—a big lake, for example—release this energy all at once. Many of these sites around the world today are open

ABOVE As a recreational lake, many areas of Fort Peck Dam are open to visitors. For example, there is a publicly accessible road running over the full length of the dam.

targets. When we think about the Fort Peck Lake scenario as described here, and other similar sites worldwide, it is important to recognize that this lake currently represents a chink in our armor. Therefore, we should eliminate the chink as quickly as possible, taking the steps necessary to protect against failure or sabotage.

If you are inclined to think that this scenario is far-fetched or fodder for aspiring terrorists, here is something for you to consider. Even though you may have never heard of this doomsday scenario, it does not mean that this idea hasn't been floating around on the internet for years. One of the first people to describe the vulnerability of Fort Peck Dam is futurist speaker and engineer Thomas Frey, who wrote about the scenario in 1988. As recently as 2017, this scenario resurfaced when Dr. Bernard Shanks, a fellow at the Resource Renewal Institute, called the Fort Peck Lake dam "the most hazardous dam in North America." The massive 2019 floods along the Missouri and Mississippi Rivers gave us a taste of what a severe Mississippi River flood can look like, echoing back to the destruction caused by the Great Flood of 1927. At one point during that flood cycle, parts of the Mississippi River became 80 miles (130 km) wide. Adding the 20-trillion-gallon pulse of water and debris from the collapse of Fort Peck Lake's dam to the natural flooding of the Mississippi River in the spring could lead to an epic catastrophe at a level never imagined before.

Here is something to consider about discussing ideas like these. Imagine that this book had come out in 1998 and that it contained the doomsday scenario that we know today as 9/11. With that warning, perhaps we would have armored all the airplane cockpit doors in 1999 and 2000. Perhaps we would have established the TSA in the same time frame, so we would have already been scanning for knives and box cutters in luggage on September 11, 2001. These measures would have prevented the terrorists behind the attack from boarding the planes with their weapons. Even if they had boarded, they would have been locked out of the cockpits. The events we witnessed on September 11, 2001, would have been impossible because the chinks in the armor that the 9/11 terrorists exploited would not have existed. The value of exploring scenarios like this is to help us see the chinks that we face as a species and as a civilization, so we can close all of them before they are ever exploited.

Please don't kill the messenger. Instead, please listen to the message.

DRONE STRIKES AND SWARMS

O n September 14, 2019, an unprecedented attack occurred on the oil refinery located in Buqayq, Saudi Arabia. Eighteen delta-wing military unmanned aerial vehicles (UAVs) and seven cruise missiles hit the refinery around 4 a.m., taking out important pieces of equipment with extreme accuracy.

Although it was portrayed in the press as a "drone attack," these were not civilian drones with four or six propellers, the kind that consumers can purchase at their local electronics superstore. The delta-wing UAVs were small military aircraft with a wingspan of about 8 feet (2.5 m) and long-range capabilities up to hundreds of miles. These drones carried small bombs. The seven cruise missiles were missile-shaped airplanes (also known as "flying torpedoes") powered by compact jet engines. They also carried bombs.

These twenty-five flying bombs hit the refinery roughly simultaneously, leading to a huge fire that took hours to extinguish. In the process, the attack completely shut down the Buqayq refinery for several weeks. This type of refinery is known as a "stabilization plant," and Buqayq's facility is the largest in the world. As crude oil comes out of the ground, it contains hydrogen sulfide and volatile organics that customers don't want or that would make the crude oil more difficult to transport on an oil tanker. For example, if the volatile organics are left in the oil on a tanker, they evaporate, creating pressure in a tank and needing to be vented off, which is not good for the environment and dangerous if they catch fire.

About 50 percent of Saudi Arabia's crude oil, and therefore 5 percent of the world's crude oil, flows through the Buqayq refinery. Bombing this particular stabilization plant with drones therefore cut off 5 percent of the global crude oil supply. In this sense, the attack was extremely effective. It disrupted the global supply chain in a significant way, causing oil prices to rise.

What this attack demonstrates is the unusual power of drones. In this case, the vehicles were small and light, making them difficult to detect. They were launched from hundreds of miles away, so, by the time they struck, their launch site was indeterminate. Their cost is extremely low compared to something like a fighter aircraft. There are no human pilots, so no one could be captured and tortured to reveal information. And most significantly, all of this technology has been democratized to the point that just about any motivated person on the planet can now afford to buy the parts and assemble their own drones, even their own cruise missiles.

RIGHT A drone attack on September 14, 2019, caused the damage to this oil processing plant at Buqayq in Saudi Arabia.

On August 23, 2019, just three weeks prior to the Buqayq attack, the movie *Angel Has Fallen* opened in theaters. The movie shows a fictional drone attack in which a swarm of approximately 100 vehicles targets the president of the United States. The drones in the movie are much smaller than the drones attacking the Buqayq refinery, but they are also delta-wing and propeller-driven.

This begs the question: what is a drone? Today, there are at least three answers:

1 There are autonomous military drones like the ones seen in Buqayq. These tend to be larger, liquid-fueled aircraft with longer ranges, and they carry military-grade explosives. In the US arsenal, the largest and deadliest of these drones are cruise missiles carrying nuclear bombs (see page 18).

2 There are remotely piloted military aircraft like the Predator drone, where the pilots are stationed in one location and the aircraft are flying in another place, sometimes thousands of miles away. The pilots control the drones using remote control via satellite links.
A Predator drone has a 50-foot (15.2 m) wingspan, costs $4 million, and can launch Hellfire missiles or similar weapons. This drone and its high-resolution cameras can stay in the air for more than 12 hours at high altitude to provide a nearly undetectable observation platform.

BELOW A UAV takes flight at the *USS George H. W. Bush*. The X-47B shown here is an example of a military drone.

3 There are remote-controlled or autonomous hobbyist drones that are now widely available to consumers. These are generally 4-rotor or 6-rotor battery-operated vehicles. At the professional level, there are 8-rotor versions with the ability to carry significant payloads. These are seen, for example, carrying Hollywood-grade cameras during movie production.

There are several reasons why even a hobbyist drone can be deadly:

- Anyone today can buy a hobbyist drone for less than $1,000, and this drone can easily carry a small explosive strapped to it. Even small drones like the ones in the movie could carry small explosives about the size of a hand grenade. And something as small as a hand grenade can do a lot of damage, especially when used on a target that is already flammable, like an airplane with thousands of gallons of fuel on board. Someone willing to spend a little more money can buy a larger drone than can carry a bomb weighing 50 pounds (22.7 kg) or more.

- It is easy to turn a consumer-grade drone into an autonomous drone by adding a small GPS module. The user can program a path of GPS coordinates, and the drone will automatically fly itself to the target without the need for any human intervention.

- Many consumer-grade drones don't have much range—they might fly about 2 miles (3.2 km) before draining the batteries. But this is more than enough range for a lot of targets. This limitation also makes the drones quite small and therefore very difficult to detect.

- It does not take that much skill for an advanced hobbyist with experience in radio-controlled airplanes to build a drone that can fly many miles. The delta-wing drones used in the Saudi Arabia attack are not that complicated or sophisticated, and a hobbyist can replicate the design fairly easily.

- It is not easy to defend against a swarm of drones, as we will see in the Prevention section. Since hobbyist drones are so inexpensive, it is easy to create a swarm.

- There are many things that a drone could carry besides a bomb, such as chemical weapons (see Chemical Attacks and Accidents, page 80) or a biological agent (see Pandemics and Biological Attacks, page 66).

- There is no human pilot involved in the mission, meaning no loss of life, no possibility of defection, much-reduced vehicle size, and improved maneuverability.

Between the fictional plot of *Angel Has Fallen* and the very real refinery attack in Saudi Arabia on September 14, 2019, we can see two ways that drone attacks unfold, one easy to imagine with a little suspension of disbelief and one that was entirely too

AI DRONES

In *Angel Has Fallen,* the drones have facial-recognition capability, allowing them to target a specific person. These recognition capabilities are a slight exaggeration on the part of the movie makers (for example, the camera resolution from a distance seems too good to be true). But the movie's depiction of the technology isn't very far off the mark. Companies like Amazon, IBM, and Microsoft have started building advanced AI capabilities like these into their cloud platforms (AWS, Watson, and Azure, respectively) so that any programmer can now access the technology easily. Creating an assassination drone (or swarm) specific to a certain person is within the realm of possibility with today's technology.

Other AI-enhanced possibilities include flying close to an airplane's engine or cockpit before exploding; targeting a specific high-value piece of equipment in a refinery, as seen in Buqayq; seeking out a dense crowd of people before exploding or releasing a chemical agent; flying near or through a specific window in a skyscraper to cause maximum damage; and flying low through the middle of a dense forest or urban environment to get to a target undetectably.

real for comfort. But are drone attacks capable of creating a doomsday scenario? Absolutely.

For example, what if the Buqayq attack had happened simultaneously alongside other coordinated attacks on multiple large refineries, pipelines, and tankers around the world? By selecting targets carefully for maximum impact, half the world's fuel infrastructure could be taken offline for a month or two before it could be repaired, causing a giant problem for the entire world economy. Drones could also attack multiple chokepoints in a nation's power grid in another possible scenario that could cripple a nation for a period of months. As portrayed in the movie, drones can take out an important world leader, leading to diplomacy problems or possibly a full-on world war scenario. This is what the United States did with the assassination of Qasem Soleimani in Iraq. If drones are used to carry out a chemical or biological attack, they could potentially kill millions of people.

Drones give a great deal of destructive power to small groups and dedicated individuals. From infrastructure attacks like those seen in Saudi Arabia to the drone swarms like those seen in the movie *Angel Has Fallen*, this technology can open a Pandora's box. Virtually anything can be attacked: refineries, pipelines, power grids, airports, fuel-tank farms, skyscrapers, schools, stadiums. Even individual people, buildings, or vehicles are vulnerable. Think about how much damage car bombs, truck bombs, and suicide bombers have done. Drones and drone swarms amp up these possibilities significantly. The fact that drones are easily accessible to millions of technically competent individuals around the world multiplies the doomsday possibilities.

ROBOT WARFARE

A drone is essentially a robot that flies. Unfortunately, this is not the only doomsday scenario that involves robots. For example, imagine a government that uses ground-based robots and drones as a police and military presence to control an entire population or to invade other countries. The US military has clearly stated its intention to remove human soldiers from the battlefield and replace them with robots. If the United States, China, Russia, and other countries were to build a hundred thousand or a million armed robots to invade another country, the invaded country would have little recourse unless it had an opposing robot army of greater strength.

Everything is in place for robot soldiers to be deployed. The United States already has aerial drones flying around armed with Hellfire missiles. Making these drones autonomous, and programming them with some relatively simple recognition software, would make them capable of finding and striking enemy soldiers and vehicles.

Countries also don't have much of an incentive to suppress the development of robot soldiers, because robot soldiers save human lives (at least for the countries that deploy them). China especially, with its massive industrial base, could deploy robot soldiers by the millions. If China or another country wanted to be aggressive, why not build millions of robots and the boats and airplanes to move them around and start invading other countries? Or let millions of robots on the ground radiate out from China and take over everything in their path? It is easy to imagine the possibilities.

There is a way to stop this escalation in robotic soldiers: diplomacy, treaties, and appeals to human decency. So far, we have been able to keep weapons out of space with treaties. For example, the Outer Space Treaty, first signed in 1967, has worked so far to prevent an arms race and war in space. Perhaps the same approach would work for robot soldiers. Maybe the strange logic of MAD (mutually assured destruction) can keep things under control, as it has so far with nuclear weapons. But keep in mind that MAD is an extremely expensive approach (both the United States and Russia spend billions on their nuclear deterrents), and it does not deter North Korea from building its own nuclear weapons and missile systems.

BELOW Robot soldiers offer several advantages over human soldiers. For example, they can be mass-produced and work 24/7 without need for food or sleep.

Modern cruise missiles were invented by the United States and first appeared in the 1980s. At the time of their debut, they truly were "miracles of modern technology." They had three features that distinguished them from any other weapon at the time:

- They can fly extremely long distances. Traveling 500 miles (800 km) is no problem for a cruise missile. It would easily be possible to launch a missile from the suburbs of Paris and accurately hit a target in London an hour later.

- Cruise missiles have the capability to map terrain very accurately, allowing them to hug the ground even in hilly or mountainous areas and avoid radar detection. This means that they can fly at 500 miles per hour (800 km/h) only 100 or 200 feet (30 or 60 m) off the ground, with the ability to change altitude second by second to handle hills, buildings, or anything else that gets in the way.

- Cruise missiles are highly accurate, and some can even visually recognize a building or vehicle before impact.

These features are relatively easy for hobbyists to replicate today because they can now buy GPS modules and small turbofan jet engines and easily access maps with terrain data and image data for much of the planet. Therefore, a hobbyist can build a pretty good facsimile of a military cruise missile today using off-the-shelf parts and some clever programming. The same goes for large delta-wing propeller-driven drones.

Consumer drones as we know them today are a more recent innovation. One of the first popular consumer models was the Parrot AR Drone, a $300 quadcopter released by Brookstone for the 2009 Christmas shopping season (and featured on its catalog cover). This marked the first time that consumers could buy an easy-to-use drone package for a reasonable price. Drones subsequently became an extremely popular and very diverse product category. Today, you can find little toy drones for less than $30, some that fit in the palm of your hand, or 8-rotor drones with payload capacities of 100 or more pounds (45 kg) if you have $10,000 or more to spend. The idea of an "affordable" passenger drone that can transport people will certainly come to fruition any day now.

What caused drones to become so popular? A set of technological innovations occurred, and these innovations were paired with some clever software. The technologies include:

- **Neodymium magnet motors:** Other types of motors were underpowered and too heavy for drones.

- **Small, lightweight, solid-state motor controllers:** These allow users to change motor power precisely and almost instantly.

- **Lightweight, powerful lithium polymer battery packs:** Other types of batteries do not provide enough power and are too heavy for drones.

- **Inexpensive 2.4 GHz spread spectrum R/C controllers:** These allows users to remotely control their drones.

- **Small, light, fast, low-power embedded processor chips:** These are used to precisely control the drone's motors and maintain stable flight.

HOW DOES A QUADCOPTER FLY?

The Wright brothers first flew their airplane in 1903, and since then airplanes have followed a pretty standard format, consisting of a tube with wings and one to four liquid-fuel engines. With its single main rotor plus tail rotor, the helicopter has a standard format, too. Invented in 1939, it has remained largely unchanged. Then the hobbyist drone in the form of an electric quadcopter appeared and threw out this century-old orthodoxy completely.

The key to quadcopter flight involves three things: the ability to precisely control the four motors so the quadcopter platform can stay level in the air despite wind and turbulence, the ability to increase or decrease power to all four motors simultaneously to change altitude while keeping the platform stable, and the ability to increase power to any two motors to move laterally in any direction. Once microprocessors were small enough and powerful enough to perform these calculations and then combined with tiny gyroscope and accelerometer chips for attitude sensing, the quadcopter drone we know and love today was born.

- **Inexpensive tiny gyroscopes and accelerometer chips:** These chips help stabilize the drone in flight.
- **Software:** New software creates stable flight for drones with four or more propellers.
- **Small, inexpensive GPS receiver chips:** These allow a drone to autonomously fly to a target.
- **Small, inexpensive 2K and 4K cameras and gimbal systems:** These features turn a drone into a high-resolution mobile camera platform.

Until all of these things came together, drones as we know them today were not really possible. These technologies, along with the power of creative entrepreneurship, combined to create the drone marketplace that we see today. It is an amazing example that shows how technological advancement in the scientific and engineering realms can enable new inventions and new product categories. Both the general public and people working in myriad commercial sectors have found many different and

ABOVE Many drones have sophisticated cameras that can capture high-resolution images.

diverse applications for drones. Farmers, emergency responders, movie producers, police officers, news organizations, real estate professionals, pipeline and powerline operators, facilities managers, security services, and Instagram influencers are just a few of the professionals benefiting from drones today.

Unfortunately, drones can also be used in nefarious ways, from drug smuggling and prison infiltration to bomb delivery. In 2018, two small drones were used in an attempt to assassinate Venezuelan president Nicolas Maduro. This is thought to be the first drone attack on a world leader.

The case of the Saudi oil refinery attack demonstrates the worst-case scenario for a drone incident:

- The drones arrived at 4 a.m.
- The swarm of drones arrived roughly at once, all flying at high speed.
- The drones flew very close to the terrain, flying "under the radar" to avoid detection.
- The drones carried high-powered bombs.
- The refinery, like most facilities today, was caught off-guard, with no practical defense system in place to protect itself against this drone swarm.

An effective defense system against an attack like this has to be able to rapidly detect and track a potentially large number of incoming drones, which may be no bigger than a foot or two (30 or 60 cm) long, and disable all the drones almost instantly so they cannot make it to the target. Since the drones may not be detected until they are quite close, the ability to react instantaneously is important.

What are the possibilities?

- **Kinetic systems:** In a kinetic system, kinetic energy in a projectile is used to destroy things. Anything that shoots pieces of metal at something else is a kinetic system. If you shoot a drone down with a shotgun, you are using a "kinetic system" to disable the drone. The Navy uses a system called Phalanx to do this kind of thing at sea with a robotic Gatling gun. Kinetic systems are more problematic on land unless there is a miles-deep buffer zone for bullets to fall into; at sea, the many pieces of flying metal that do not hit a target fall harmlessly into the ocean. Slewing the gun accurately a hundred times in a few seconds to eliminate a drone swarm is also difficult, so now there needs to be multiple guns. Think about it this way—if you have a hundred drones incoming, one gun, and ten seconds to react, the gun has to move, lock on a target, and fire a hundred times in those ten seconds to hit all of the drones. Since that is unlikely for a single gun, now we need ten guns, or fifty. Because the realistic range of a gun is half a mile (0.8 km) or a mile (1.6 km) at best, we would need multiple gun arrays to cover a large facility.

- **Net systems:** Using a net-shooting gun to throw a net at a drone can entangle and disable it. It is a little hard to imagine throwing a hundred nets accurately in a small period of time, but shooting canisters that release nets at a rapid-fire clip could have potential if there are enough guns. A net-shooting gun has the same slewing problem as the kinetic system. One advantage of nets over bullets is that falling nets tend to be less harmful to people and wildlife on the ground.

- **Directed energy beams and lasers:** It is possible to shoot high-energy, focused microwave beams at drones and disable their electronics, provided that the electronics are not shielded. Lasers work similarly, and they disable the electronics by melting a hole in the target instead. Again, the beam has to slew, track, and hit a hundred or more targets very quickly. The

CHAMP system (Counter-Electronics High Power Microwave Advanced Missile Project) is one example of a radiofrequency energy weapon like this, as well as Israel's Iron Beam system and Drone Dome.

- **EMP counterattack:** If the drones are not adequately shielded, a small EMP weapon (see page 44) with a mile (1.6 km) radius can disable their electronics with an electromagnetic pulse. Of course, the downside to this is that everything else electronic inside the radius would be disabled as well if it lacks shielding.

- **Drone-on-drone hunting:** A counter-swarm of drones numbering in the hundreds is launched to attack the incoming drones. These defensive counter-drones would need to be extremely nimble and accurate and either run into or shoot the incoming drones to disable them.

The problem with these defense systems is that they would need to be deployed nearly everywhere. This becomes difficult and expensive to implement. Such systems can conceivably be deployed to protect high-value, compact targets like refineries, stadiums, and airports. But protecting a pipeline or a power grid that is a thousand miles long could become extremely expensive. Protecting every school, university, skyscraper, business, factory, chemical plant, and large event venue would be costly too.

There is no question that the threat is real. At this moment, there is not a good answer to that threat, but here are some possibilities:

- Provide security measures when gathering large crowds of people together, especially in outdoor venues. There is no question that gathering 50,000 or 100,000 people together in one compact place makes these venues an obvious target.

- Prioritize the entrepreneurial development of low-cost drone defense systems, so that these systems become as ubiquitous and inexpensive as chain-link fences.

- Make drones illegal, although it is easy to suspect that this will not work. For example, heroin is illegal yet widely available in the United States (and most other countries).

- Deploy municipal drone swarms in the same way we deploy police cars, blanketing each city's skies with drones that shoot down (or disable) everything that flies.

- Use some form of universal spying and intelligence gathering (see page 88) to attempt to detect bad actors while they are in the planning stages.

One problem with drones is that they give just about anybody the power to attack just about anything. A second problem with drones is that defending against them tends to be difficult and expensive with current technology. Drone swarms amplify these two problems. It will take new, creative solutions to minimize this threat.

ABOVE A Phalanx CIWS unit on the USS *Nimitz* fires a shot during a calibration exercise.

NUCLEAR BOMBS

THREAT LEVEL: WORLD

You are working away in your typical corporate office in New York City. You are sitting in a low-walled cubicle when you hear your phone screech with an alert. You soon hear nearby phones from your co-workers start screeching as well. Then you hear several people scream. You see people leap up from their chairs and start running for the door. You open the alert on your phone, and it says: "BALLISTIC MISSILE THREAT INBOUND TO NEW YORK. SEEK IMMEDIATE SHELTER. THIS IS NOT A DRILL."

For a moment, step back and think about how you might react. You have received a completely credible warning on your phone's special emergency alert system about an incoming nuclear missile. Yes, it says "ballistic missile," but that means a nuke. You may not know exactly how nuclear missiles work, how bad the explosion might be, or how long it takes for a ballistic missile to arrive, but you have seen the pictures and heard about all of the death at Hiroshima. You know, for certain, that you are about to die. You are holding in your hand a message indicating that your death is imminent, probably only minutes away. And so does everyone else around you. What is your next move? How would you react?

This scenario actually happened on Saturday, January 13, 2018, although in Hawaii rather than New York. The alert went out to phones and also appeared on TV and on radio stations, just as any emergency alert would.

At the time this happened, tensions with North Korea were high. North Korea had been actively testing both nuclear bombs and ballistic missiles to deliver them. There was relentless news coverage about the North Korean threat at the time, so the idea that North Korea might strike Hawaii with a nuclear bomb was on everyone's mind. As a result, many recipients of the alert unequivocally interpreted it as a real warning about an impending death sentence. It took thirty-eight minutes for the state to rescind the alert, and in those thirty-eight minutes, people had good reason to believe they were about to die.

If we were living in the 1950s, we would know something about what to do. The threat of an imminent nuclear strike from the Soviet Union was a major concern during the Cold War. Students in school performed "duck and cover" drills, homeowners built their own underground fallout shelters, and any big building with a basement had a highly visible

RIGHT This mushroom-shaped cloud came from the atomic bomb that was detonated over Nagasaki on August 9, 1945.

22

Civil Defense sign marking it as an emergency fallout shelter, with barrels of supplies on hand to keep people alive in the aftermath of the disaster.

Today, we, as a society, have none of this groundwork in place. So, how did people in Hawaii react? Their responses came in several flavors:

- Many people tried to call loved ones to say good-bye. As in the September 11 attacks, the number of calls saturated the phone system, meaning many calls were blocked. So many people called 9-1-1 services that they, too, were overwhelmed.

- Others tried to drive to loved ones so they could physically be with them during the end. Parents, for example, drove toward homes or schools to be with their children. There were reports of cars barreling down the road at 100 miles per hour (160 km/h).

- Many people sought shelter, as advised. They moved toward basements or concrete buildings. Some holed up in caves. Many in cars pulled over and parked their cars in tunnels, seeking any kind of cover from the imminent explosion.

Perhaps the most interesting reaction came from a short video clip showing parents lowering children into storm sewers. This clip received widespread media attention because it is so heart-wrenching. We can easily imagine what is going through the parent's mind and the child's.

To believe, without doubt, that you are about to die within minutes would be a life-changing event for a lot of people. And then to be rescued thirty-eight minutes later, deus-ex-machina style, by the fact that it was a false alarm carries some weight as well. The sense of relief would be overwhelming. Many were also furious that they had to face an imaginary existential crisis that day.

Now, let's take this scenario to the next level. What if the alert had been true? What if a missile had actually struck and obliterated Honolulu in the same way Hiroshima was obliterated in 1945? Or what if we found that 100 missiles had been launched toward the continental United States? Here we would be talking about a nuclear holocaust, with tens of millions of people dead, the economic machinery of an entire country destroyed, and then a very high probability for a nuclear winter that engulfs the planet.

SCENARIO

Around the world today, at least eight countries have nuclear weapons. The United States has an arsenal of approximately 6,000 nuclear bombs. At any given time, approximately 1,800 of these are mounted on missiles that are ready to fly, and there are other types of nuclear bombs as well. Russia has roughly the same. China, France, and England have much smaller arsenals, more like 300 bombs. And then there are countries like North Korea, India, and Pakistan with 200 bombs or less.

One typical bomb in the US arsenal is the B83. The United States has built over 600 of these. It is a 1.2-megaton thermonuclear weapon. This means that when it detonates, it explodes with the power of 2.4 billion pounds (1.1 billion kg) of TNT. This is an unimaginably large amount. To get an idea, picture a cube that has sides that measure approximately 300 feet (91.5 m). Or a 30-story building shaped like a cube made from TNT. A bomb like this is about a hundred times more powerful than the bomb dropped on Hiroshima.

There are three scenarios that arise from nuclear missiles, depending on their targets and how many missiles are launched:

1 **One or two nuclear missiles strike one or two minor targets, like Honolulu.** This is not to say that a strike on Honolulu is not catastrophic, but Honolulu is a comparatively small city and not strategically important to national interests.

2 **One or two nuclear missiles strike one or two major targets like London and New York City.** The destruction of major cities like these would severely impact the world economy.

3 **A large volley of nuclear missiles (100 or more) is fired, striking targets in several countries through attacks and then counterattacks.** In this case, we have to worry about global effects from a phenomenon called nuclear winter.

In any of these scenarios, what happens when a modern nuclear bomb strikes a target like a city? Its effects will depend on the distance from the detonation and the size of the bomb. Assuming that a B83 bomb is detonated in the air, as with the bomb in Hiroshima, we would see the following:

- In the "nuclear fireball" zone, a sphere that surrounds the detonation about 1.6 miles (2.5 km) in diameter, everything is immediately and utterly destroyed.

- In the second ring, called the 5 PSI (pounds per square inch) air-blast area, shock waves of compressed air radiate out from the explosion.

This circle is about 9.3 miles (15 km) in diameter. A big, sturdy concrete building might survive, but wood frame houses would collapse. Anyone outside, or anyone inside a collapsing building, would probably die.

- The next circle, known as the thermal radiation zone, is 16 miles (26 km) in diameter. When the bomb explodes, it creates a tremendous amount of heat and light. The intense heat ignites anything flammable in this zone and instantly burns anyone outside with third-degree burns. Some pictures from Hiroshima show a shadow of a person on an exterior wall or sidewalk; these shadows were created when a person's body blocked the intense light and heat from the bomb, which bleached or burned the uncovered areas.

- The outer circle is known as the 1 PSI air pressure zone and is 26 miles (42 km) in diameter. All the windows in this zone are likely to shatter from the blast wave. Anyone in this zone would see a flash of light, and if they turned to look out a nearby window, the window would spray shattered glass in their face a bit later.

The "advantage" of an air burst is that there is minimal fallout, radioactive contamination on the ground, and radiation poisoning, relatively speaking. The bomb also doesn't leave a crater. This is why Hiroshima is a bustling metropolis of 1.2 million people today rather than a radioactive wasteland. If a nuclear bomb detonates on the ground instead of in the air, the explosion irradiates soil and sends it skyward, leaving a crater. When these contaminated particles settle back on the ground, it is known as fallout. Wind can spread fallout significant distances from the blast site, meaning that fallout can potentially contaminate thousands of square miles of land. In addition, people in the vicinity of the bomb are irradiated.

One way to discuss the exposure to radiation from fallout is in terms of rads per hour. To put it in perspective, a person might experience half a rad per day of radiation from normal natural sources. A chest X-ray might be at a level of 5 rads and is generally considered harmless. If a B83 bomb detonates on the ground, fallout radiation at a level of 10 rads per hour will be present in an area 110 miles (180 km) downwind from the blast site with normal wind levels.

If a nuclear airburst bomb of this size struck a major metropolitan city, we would expect to see the city destroyed, along with surrounding areas. Millions would be killed or injured, and, depending on the city, the world might lose a financial hub, important corporate headquarters, world leaders, an industry hub, and cultural relics.

What if one country wants to destroy another and launches a large volley of nuclear missiles? If 100 bombs land in the largest cities in the United States, something on the order of 50 million to 100 million people die, and the destruction will ruin so many industries and economic enterprises that the country may not recover in any reasonable time frame. For example, just a few bombs would destroy most of the ability of the United States to refine gasoline or take down most of the internet in the country.

With this many bombs, nuclear winter becomes a likely possibility. The dust and debris from the explosions, along with the soot from the fires on the ground, can contaminate the global atmosphere for a long period of time—from months to years. Sunlight has trouble penetrating the particles in the upper atmosphere, causing global temperatures to drop. The resulting cold temperatures and lack of sunlight could cause widespread famine because of crop failures.

BELOW The atomic bomb that was dropped on Hiroshima on August 6, 1945, destroyed about 70 percent of buildings in the city.

SUITCASE NUKES

Is it really possible for a nuclear bomb to fit inside a vending machine? Or the trunk of a car? Or even a proverbial suitcase? What about a backpack that lets someone easily walk around an urban area undetected with a nuclear bomb strapped on their back? The answer is yes.

A portable atomic bomb—known as a suitcase nuclear device—is a miracle of miniaturization. In the same way that many devices such as cell phones and computers shrank from their rather large original sizes as technology advanced, so too have nuclear bombs.

The bomb dropped on Hiroshima was enormous and weighed 9,000 pounds. But both the United States and Russia have developed ways to shrink atomic bombs. The motivation is simple: a desire to fit multiple nuclear warheads on a single ballistic missile, or to fit them inside a cruise missile, or to fit them inside the form factor of a standard artillery shell, or to create bunker-buster bombs.

A suitcase nuclear device is tiny, measuring just 16 inches (40 cm) long and 11 inches (28 cm) in diameter and weighing only 50 pounds (23 kg). How is it possible to make a nuclear bomb this small? The answer involves plutonium, along with some clever engineering. Let's assume that the device has a softball-size plutonium core. Plutonium is a very dense metal (similar to gold in its density), so a core of this size weighs about 14 pounds (6 kg).

At 14 pounds (6 kg), there isn't enough plutonium for the metal to explode, as there are not enough spontaneous neutrons being generated in order for a chain reaction to start (see page 29). In order to cause this sphere of plutonium to "go critical" and explode, several techniques could be used:

- A substance known to generate neutrons could be added to the mix. For example, the plutonium sphere could hold a much smaller polonium-beryllium sphere in its center. This smaller sphere is known as an initiator, and it triggers off a burst of neutrons once it's activated.

- A tamper, or neutron reflector, can be arranged around the core so that the neutrons that are naturally escaping from the sphere are reflected back inside.

- High explosives can compress the plutonium sphere to a greater density than normal, so that the smaller mass becomes critical.

In the case of a miniaturized atomic bomb, all three techniques come together to create a nuclear explosion from a small package. A bomb this size can easily destroy an area as large as a city block. There are many potential targets, especially when a bomber can transport the bomb in a wheeled suitcase or a backpack. If the blast occurs slightly underground rather than in the air, there would also be significant fallout and ground contamination that could make the city, along with nearby areas underneath the plume, poisonous for decades.

Given that this doomsday scenario would be incredibly destructive, why hasn't it happened yet? Only a large organization such as a nation-state can create plutonium. Plutonium, being a synthetic (man-made) element, cannot be found in nature, and access to a nuclear reactor is required to create it. Then you need elaborate equipment to isolate and purify and shape the plutonium. The good news: no one is going to be creating plutonium cores in their basement.

The source of power for all nuclear weapons (along with nuclear power plants) starts with a property found in several elements in the periodic table. Two of these elements—uranium-235 (U-235) and plutonium-239 (P-239)—form the foundation of the nuclear-bomb industry.

Let's start at the beginning. The first thing to understand is that most elements in the periodic table that we encounter in our everyday lives are stable. For example, most carbon atoms in a piece of charcoal or a carbon-fiber tennis racquet are carbon-12 atoms. These atoms have six protons and six neutrons in their nucleus. If we put a trillion carbon-12 atoms in a jar, and then open the jar 100,000 years from now, it will contain a trillion carbon-12 atoms. (For sticklers, one trillion carbon atoms constitute a tiny amount of carbon. Just 1 gram of carbon-12 in reality contains 50 billion trillion atoms of carbon-12, so one trillion atoms would be one fifty-billionth of a gram. The number "trillion" here is used as a familiar placeholder.)

But there is another form, or isotope, of carbon called carbon-14. It has six protons like carbon 12 does, but eight neutrons instead of the six found in carbon-12. Carbon-14 is not stable; instead, it is a radioactive element. If we put a trillion carbon-14 atoms in a jar and come back to open the jar 100,000 years from now, the jar will mostly contain nitrogen gas (nitrogen-14, in this case), with only about 8 million carbon-14 atoms left.

What happened? Why did the carbon-14 atoms disappear and turn into nitrogen-14? Because carbon-14 is radioactive, it undergoes a process called beta decay, meaning that a surplus neutron in a carbon-14 atom can spontaneously transform into a proton, an electron, and an anti-neutrino. After beta decay, a carbon-14 atom with six protons and eight neutrons turns into a nitrogen-14 atom with seven protons and seven neutrons. Carbon-14 has a half-life of 5,730 years; this means that every 5,730 years, half of the carbon-14 atoms in our jar turn into nitrogen-14 atoms. Nitrogen-14 is stable, so, over time, a jar full of carbon-14 turns into a jar of nitrogen-14 at a steady rate. If you have heard of the carbon-14 dating technique used to determine the age of wood or bone, this natural decay process is the basis of the technique.

Many isotopes of different elements in the periodic table are radioactive, and they decay in various ways. They give off alpha particles, beta particles, gamma rays, and/or neutrons depending on how they decay.

The two isotopes (U-235 and P-239) used to make nuclear bombs are radioactive. They also possess an additional property that makes them bomb-worthy: they are fissile. If we hit one of these atoms with a flying neutron, the atom will split almost instantly (in picoseconds) into two smaller atoms. For example, a U-235 atom might capture a neutron and spontaneously split into krypton-92 and barium-141; in the process, it will emit three new flying neutrons. If an amount of U-235 is arranged in an optimal shape and has a sufficient mass (known as the critical mass), all three of these new flying neutrons will likely hit three more U-235 atoms that will also split, giving off nine new flying neutrons. After just fifty iterations of this process, we have trillions of flying neutrons in less than a nanosecond. A nuclear chain reaction is well under way.

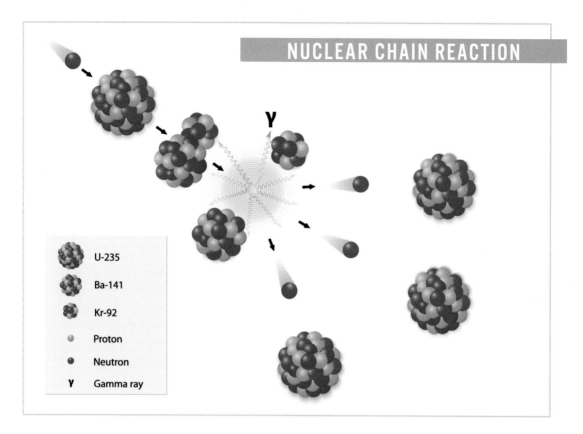

NUCLEAR CHAIN REACTION

U-235

Ba-141

Kr-92

Proton

Neutron

γ Gamma ray

ABOVE In a nuclear chain reaction, a flying neutron strikes an atom of uranium-235. In the example depicted in this diagram, it causes the atom to split into an atom of barium-141, an atom of krypton-92, and three neutrons. This collision also releases gamma radiation. The three neutrons continue the chain reaction by hitting other uranium-235 atoms.

The power of a nuclear bomb comes from another interesting property of U-235: when the atom splits, it gives off energy, primarily in the form of heat, light, and gamma rays. If a bomb designer creates a bomb configuration that causes 10 kilograms of U-235 to undergo high-speed fission, the bomb instantly generates 830 terajoules of energy. For comparison, a ton of TNT creates 4 gigajoules of energy when it explodes, so 10 kilograms of U-235 is the equivalent of 207,000 tons of TNT.

P-239 is similar to U-235 in that it is also fissile. Each time an atom of Pu-239 splits, it gives off three neutrons, meaning that a very rapid nuclear

chain reaction can occur under the right conditions. One kilogram of P-239 generates the same amount of energy (heat, light, gamma rays) as one kilogram (2.2 pounds) of U-235 during fission.

The difference between U-235 and P-239 is that it takes much less plutonium to make a bomb. This has to do with the critical mass of the element required. A sphere of U-235 would need to weigh about 120 pounds (about 55 kg). For P-239, we would need only 24 pounds (11 kg). If we have a sphere of U-235 of any size, then some of the flying neutrons will hit other U-235 atoms and others will fly outside the sphere. At critical mass, there are enough neutrons inside the sphere for the chain reaction described above to occur. Below critical mass, there aren't enough flying neutrons inside the sphere for the chain reaction to sustain itself.

Given this knowledge, it is easy to understand the mechanics behind the simplest possible nuclear bomb. We form two half-spheres of U-235 weighing 60 pounds (27 kg) each. As long as these hemispheres are separate, nothing will happen. But if we bring them together, we have created a critical mass and the bomb explodes.

This is the basic concept behind the U-235 bomb dropped at Hiroshima. In this case, rather than a sphere, the two pieces of uranium were shaped as a cylinder of uranium (imagine a soda can shape) that could fit inside a ring of uranium (like a Koozie® around a can of soda). The ring shoots down the barrel of a cannon at the cylinder, so that the two pieces of U-235 assemble rapidly. Since the combination is not perfectly spherical, they added 20 extra pounds of U-235 to make up the difference. When the ring surrounds the cylinder, the bomb explodes.

This is an extremely simple architecture, but it has two notable problems. First, the Hiroshima bomb was extremely large and heavy, because it contained over 100 pounds of uranium, the cannon, the explosives to fire the cannon, and other necessary components. Second, the bomb might only have an efficiency of 1 or 2 percent. Once the uranium starts to fly apart, it is no longer at critical mass, so the chain reaction stops. Perhaps only 1 percent of the uranium has a chance to undergo fission before the critical mass disassembles. In the Hiroshima bomb, that very small percentage still resulted in a 15-kiloton yield, but ideally the objective would be to use up a lot more of the fuel to generate a bigger explosion.

This thinking led to the more sophisticated implosion architecture. In the Nagasaki bomb, there was a 14-pound (6.4 kg) sphere of plutonium at the core—not a critical mass. Inside the core was a neutron generator that would be triggered when the bomb detonated. And then surrounding the core was a "tamper" of U-238, surrounded by a thick aluminum "pusher." And then surrounding all this was a 6,000-pound (2,720 kg) sphere of conventional explosives called "composition B" (similar to TNT, and found inside things like artillery shells and hand grenades). When triggered, the composition B layer detonates, creating a tremendous pressure wave inward. This explosion compresses the plutonium sphere enough for it to become a critical mass, and also sets off the neutron generator in the center of the plutonium sphere. Meanwhile, the U-238 sphere is heavy and compressing as well, so it has two effects: First, it reflects neutrons that would otherwise escape, and second, it holds the plutonium sphere together longer, resisting the sphere's tendency to fly apart. Therefore, this bomb is much more efficient, perhaps ten times more efficient than the Hiroshima bomb, because of the ability to hold the plutonium together in a critical-mass configuration longer.

Subsequent improvements in technology have allowed implosion-type Pu-239 bombs to shrink considerably through improvements in explosives, neutron generators, and neutron reflectors.

In addition, scientists learned how to create two-stage fusion bombs, also known as thermonuclear bombs, with greatly enhanced explosive power. In thermonuclear weapons, an implosion-type fission bomb (known as the primary fission) is used to compress and irradiate hydrogen to the point where hydrogen fusion occurs, which ignites another fission reaction (known as the secondary fission) in the container around the hydrogen. This improvement in bomb architecture allows the explosive power of thermonuclear bombs to move into the megaton range.

WHERE DOES U-235 COME FROM?

The simplest nuclear bomb starts with uranium ore, dug out of the ground. This ore is fairly abundant in a number of mining areas around the world. Once the ore is extracted, it is processed into a sand-like substance called yellowcake. Yellowcake can then be processed and refined into uranium metal.

Only a tiny percentage (0.7%) of this natural uranium metal is made up of U-235. The rest is U-238. The U-235 is extracted and purified through a process known as *enrichment*. It involves combining the uranium metal with fluorine, so the metal turns into a gas. U-238 gas molecules are slightly heavier than U-235 molecules, so the molecules are put into sophisticated high-speed centrifuges to separate the isotopes. A nearly pure form of U-235, also known as *highly enriched uranium*, can be made with centrifuges. The U-238 left over is called *depleted uranium*.

RIGHT Yellowcake (ammonium diuranate) is packaged and transported in steel drums.

PREVENTION

Humanity faces two threats from nuclear weapons:

1 The threat that comes when one nation-state uses nuclear bombs on another;

2 The threat that comes if/when terrorists get ahold of a nuclear weapon and decide to use it.

Nuclear deterrence between nation-states revolves around a theory called mutually assured destruction (MAD). The theory rests on the threat of retaliation. Any nation is kept in line by the threat that one nation's strike on another would likely result in the complete annihilation of the aggressor nation itself as retribution. The name for this concept was coined in 1962 by an analyst named Donald Brennan. And while it sounds insane, it works. In a world where two nations can utterly obliterate each other with nuclear weapons, neither side has motivation to use said weapons for fear of this obliteration.

In this landscape, is there any way to prevent a nuclear attack via ballistic missiles besides the threat of retaliation? One strategy falls under the colloquial name of the "Star Wars defense" and is officially known as the "Strategic Defense Initiative." With this strategy, a nation tries to develop defensive weapons that can neutralize any potential missile strike. There would be an attempt to disable missiles as they launch, as they are firing their engines to reach orbit, as they reach space and coast, or as they re-enter and fall toward targets on Earth.

The first two stages of the missile strike are fairly easy to detect using a satellite. When the missile reaches space and begins falling toward Earth, detection becomes more difficult, but it is possible using radar. The last phase is the most difficult to deter, because the enemy can employ MIRV (multiple independent re-entry vehicle) technology to deliver many warheads with one missile and can also use decoys and distractions (such as chaff or flares) to make the warheads harder to target.

The United States' Patriot Missile Long-Range Air-Defense system is one example of how a nation can intercept a warhead as it is falling to Earth. The Patriot missile launches from the ground and rams into the warhead (similar to a bullet hitting a bullet), destroying the warhead with kinetic energy and a small explosive. The warhead is first detected by a ground radar system, and then the missile's internal radar takes over once it approaches the target warhead. THAAD (Terminal High Altitude Area Defense) is a similar defense system that uses two-stage rockets.

Because a THAAD missile hits a warhead at a higher altitude, it can protect a larger area. The Aegis Missile Defense System has an even longer range.

There are a few problems with missile defense systems like these. One issue is the number of missiles needed to handle one incoming intercontinental ballistic missile (ICBM). We need as many as fifteen defensive missiles to deal with one incoming ballistic missile if we want to handle all of the MIRV warheads along with any decoys. Another issue is the limited range of some missiles. For example, the United States would need Patriot missile batteries in all major cities in order to protect them.

While these defense systems can help create a shield against ballistic missiles, nuclear bombs do not have to arrive in this way. They can be transported via a cruise missile, which is much harder to detect because it flies like an airplane and very low to the ground. In the worst-case scenario, a submarine near shore would launch a cruise missile, reducing its flight time to mere minutes and making interception much more

ABOVE Visitors to the Titan Missile Museum in Green Valley, Arizona, can see a preserved *Titan II* missile site. The inert intercontinental ballistic missile is shown loaded into a silo.

ABOVE During the Cold War, some communities in the United States built fallout shelters as a civil defense measure against a nuclear attack from the Soviet Union. This drawing shows a fallout shelter built beneath an elevated highway.

difficult. The prevention strategy here is to install radar systems in the air so that they can look down and detect low-flying craft like cruise missiles. Once the radar detects the threat, it would then need to instantly launch a missile to destroy the cruise missile.

Several decades ago, there was a focus on preparing for attacks with duck-and-cover drills, mass evacuations of cities, and civil defense shelters in building basements. But today, where a ballistic missile can arrive in an hour or less, evacuation does not make sense (especially at the scale of today's cities). This is why the discussion has focused instead on missile defense systems. To prepare people, we might train them to stay away from glass windows after seeing a bright flash. In the case of both nuclear bombs and an asteroid strike (see page 122), the bright flash is followed by a blast wave that, among other things, shatters windows and shoots glass fragments at anyone looking out. We could also educate people on what to do in the aftermath of a bomb. The best thing to do is to get inside, preferably in the basement or interior

of a concrete building, and stay there for 24 hours so the peak radiation can dissipate.

In an ideal world, another method for avoiding nuclear warfare would be to destroy all existing nuclear weapons, along with all nuclear reactors and nuclear centrifuges on the planet. But this seems unlikely. Throughout history, humans have had a hard time getting along. There is always the risk that a rogue nation could secretly create new reactors and new bombs, then hold the rest of the world hostage. Humanity may also need nuclear weapons to avoid another doomsday scenario, for example, to change the path of an asteroid (see page 131) or to destroy an alien mothership (see page 234).

So for now we are left with MAD, diplomacy, advancing technologies in the missile-defense realm, and the goal of preventing any more nations (or terrorists) from gaining the ability to create nuclear bombs. Though this is unfair to nations without nuclear arms because they have no retaliatory power, it may be the best we can do.

ANTARCTICA COLLAPSE

THREAT LEVEL: WORLD

E veryone started paying attention when the news came out about the bomb in Antarctica. The story that unfolded was both horrific and diabolical. A splinter group of a well-known terrorist organization had claimed responsibility for the bomb, delivering a long prerecorded speech on their reasoning and methods. According to their video, they had acquired one high-yield nuclear weapon, just one, through an intricate terror network. They wanted to deploy it in order to cause maximum pain and destruction around the world, so they decided to detonate it in Antarctica.

Detonating the bomb in a major city would cause damage, yes, but their fear was that the damage would be superficial—that it was possible for a nation to quickly recover from the loss of just one city. The terrorists wanted a global impact, real revenge, and a lasting message that could not be reversed.

What they had accomplished was an immense chain reaction in West Antarctica. The nuclear bomb destabilized an enormous amount of ice—millions of cubic miles of ice. The ice in West Antarctica had already become relatively unstable, so the single bomb simply helped that process along. Over the course of several months, one thing led to another, and more and more of the destabilized ice fell into the sea. Aided by unusually high summer temperatures, the collapsing ice in one area destabilized ice in other areas until ocean levels eventually rose by about 8 feet (2.5 m) around the world. Everyone had expected sea levels to rise, but slowly, over centuries, as Antarctic ice melted due to climate change. Now it had happened in essentially an instant, with the world completely unprepared.

Miami was completely lost. In Washington, DC, the parts of the city near sea level were underwater. Flooding destroyed large parts of San Francisco, Boston, and New York City. Perhaps half of Bangladesh, a nation with a population of 160 million, was flooded, as were half of the Netherlands, urban centers like Shanghai and Venice, all kinds of low-lying islands and archipelagos, and thousands of smaller coastal and beach towns around the world. The Thames River, which flows from London to the sea, saw many of its banks flooded out. The official elevation of London, New York City, and Washington, DC above sea level had been 36 feet, 33 feet, and 13 feet (11, 10, and 4 m), respectively,

RIGHT This series of satellite images shows the collapse of the Larsen B Ice Shelf between January and March 2002. Notice how the shelf collapses all the way back to the hills on the left and is replaced with a light blue slush. The destabilization of ice in other parts of Antarctica, especially West Antarctica, may lead to the collapse of other ice shelves in the continent.

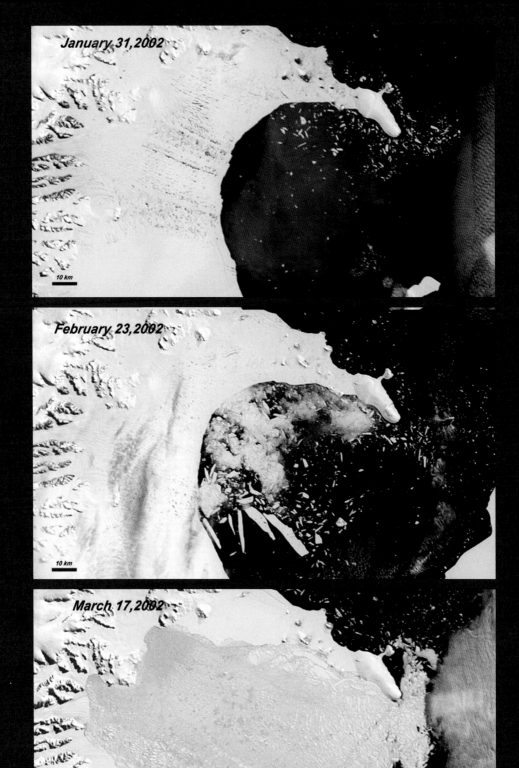

so an additional 8 feet of water added to the planet's sea level was a huge game-changer. The terrorists had succeeded beyond their wildest dreams.

There were immediate worries about the additional side effects on the climate. With the ocean so much higher, coastal stormwaters from hurricanes or typhoons could now reach much farther inland. Combine a big storm surge and high tides, and the results in low-lying cities were catastrophic. Some of the ice in East Antarctica were now becoming unstable as well, making the potential problems even worse.

One bomb in the right place had changed the world in an instant. Many cities and hundreds of millions of people saw their futures altered forever. The loss of property, housing, urban infrastructure, and industries like tourism was shocking. The number of sea-level refugees who had lost everything with nowhere to go was extreme.

SCENARIO

Antarctica is divided into two parts: East Antarctica and West Antarctica. West Antarctica holds enough ice to raise global sea levels about 10 feet (3 m). East Antarctica holds enough ice to raise global sea levels by close to 200 feet (61 m). While East Antarctica is stable at the moment, West Antarctica is a different story. Scientists are already talking seriously about the natural collapse of West Antarctica glaciers and ice shelves. While it might take a thousand years for East Antarctica to completely melt, West Antarctica is much more dynamic, though not in an instantaneous sense. This means that the changes will happen within decades, rather than centuries.

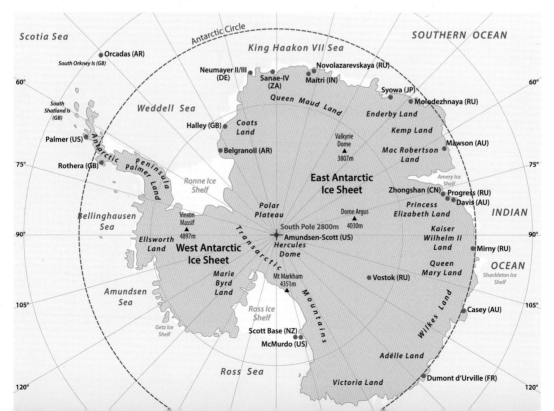

West Antarctica has several glaciers that travel in different directions toward the ocean. One of these glaciers, the Thwaites Glacier, is already known as the "doomsday glacier." It earned this nickname for several reasons. It is enormous—similar in size to Great Britain and more than a mile thick—and it is already moving quickly (for a glacier) at about 1.25 miles per year (2 km/yr). If it collapsed, adjacent glaciers would collapse as well, destabilizing the region and potentially leading much of West Antarctica's ice to crumble into the sea. For example, the Pine Island Glacier is immediately adjacent to Thwaites, is nearly as big, and is also moving fast.

A doomsday scenario involving the Thwaites Glacier can unfold in two ways. If a rogue terrorist group were to detonate a nuclear weapon in Antarctica, the bomb could have the power to quickly destabilize the whole region, leading to a rapid collapse of the glaciers in West Antarctica and a rise in sea levels around the globe. As described in the beginning of this chapter, sea levels might rise 8 feet (2,5 m) in a year.

The collapse of this glacier can also happen naturally, albeit more slowly. As we'll see in the Science section, the precarious nature of the ice in West Antarctica is a big problem for the planet, even without an accelerant like a bomb. As impossible as it sounds, Antarctica had its first heat wave in 2020, with a new high temperature approaching 70°F (21°C). Ice is melting in Antarctica much faster than expected, which only adds to the instability seen in West Antarctica.

In either case the situation is dire, given the number of coastal cities that could be affected. It would be possible to evacuate cities before they are submerged, so loss of life is not an issue. The problem is a gigantic investment in buildings and infrastructure that would be lost. In Manhattan, the Freedom Tower alone cost $4 billion to build. There is also a cost for individuals who live in these cities. Many people who will be displaced when these cities go underwater will lose their homes, their property, and their livelihoods. More than 100 million people would become refugees, never mind the immense financial hit from the buildings, structures, and infrastructure.

Take Shanghai, China, population 24 million, as a simple example. Its average elevation is 13 feet (4 m) above sea level. If the West Antarctica ice collapsed and caused sea levels to rise as described, much of Shanghai would be lost.

LEFT This map of Antarctica shows its two major regions and the research stations scattered throughout the continent.
RIGHT In October 2018, unusually strong high tides and storms led to major flooding in Venice. The floodwaters peaked at about 6 feet (1.87 m), the highest level the city had seen in fifty years.

WHAT'S THE DIFFERENCE BETWEEN A GLACIER AND AN ICE SHEET?

A *glacier* can be thought of as a slow-moving (e.g., 1 mile, or 1.6 km, per year) river of ice that typically forms in mountainous regions and moves downhill. There are many glaciers all over the planet. The best-known ones are in the Himalayas, several of which are around Mt. Everest. The United States even has a glacier national park in the Rocky Mountains.

An *ice sheet* is a gigantic mass of ice. There are only two on the planet, one in Greenland and the other in Antarctica. The Greenland ice sheet is 660,000 square miles (1,700,000 km²) in size and more than a mile (1.6 km) thick. The Antarctic ice sheet is roughly ten times larger and contains on the order of 53 quintillion pounds, or 53 million trillion pounds (24 quintillion kg) of ice.

One other term that often enters the conversation is an *ice shelf*. When a glacier hits an ocean, the ice of the glacier starts floating and can often extend quite a way out from land. The biggest ice shelf in Antarctica today is called the Ross ice shelf, fed by multiple glaciers. It is an enormous floating mass of ice roughly 500 x 400 miles (800 x 650 km) in size. The glaciers are adding to the body of the ice shelf every day, while icebergs calve off of the ice shelf's edges far out in the water.

BELOW Located in Montana, Glacier National Park offers panoramic views of valleys and mountains that were carved by glaciers and the glaciers themselves. It is predicted that all the glaciers will be gone within a few decades.

Looking at it from a scientific standpoint, Antarctica is a fascinating place. Given its location at the Earth's South Pole, Antarctica receives very little heat from the sun compared to, for example, areas along the equator. Antarctica therefore should be extremely cold—its icy nature is natural.

Historically, the coldest months in Antarctica are June, July, and August, when the average daily temperatures at are –13°F (–25°C) at McMurdo station and –85 degrees F (–65°C) at Vostok station. However, in the warmer months recently, temperatures have been unnaturally balmy. The highest summertime temperature recorded so far is 69.35°F (20.75°C). Under these warmer conditions, summertime melting in Antarctica is accelerating as the Earth's atmosphere and oceans warm.

Why is the Thwaites Glacier so critical and so worrisome? Why is it considered to be a "linchpin" that could trigger the collapse of much of West Antarctica?

The main body of the glacier rests on Antarctica's submerged continental shelf, the largely invisible land that lies beneath all of Antarctica's ice. The Thwaites glacier naturally slides toward the ocean as

BELOW Melting in six glaciers in West Antarctica has accelerated since the 1990s. The red indicates areas that have seen increases in the velocity that these glaciers travel.

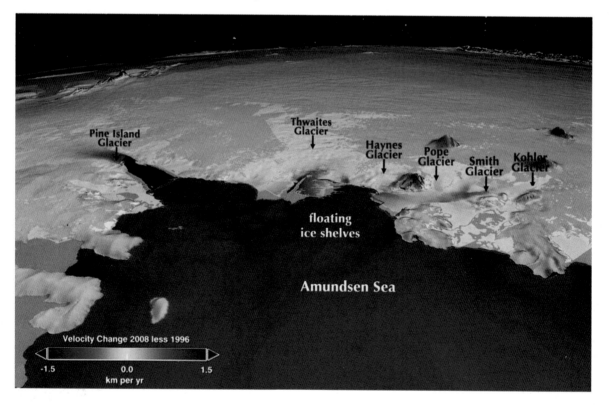

ANTARCTICA COLLAPSE

gravity pulls the glacier downhill. A floating ice shelf forms when the glacier reaches the water. Icebergs naturally break off from the ice shelf as the end of the shelf thins and weakens.

In a scenario where the glacier rapidly collapses, the floating part of the glacier—the ice shelf—breaks away much more quickly than normal. With the ice shelf gone, the body of the glacier becomes exposed. This exposed section makes up the glacier's ice face, and it now towers a quarter of a mile (400 m) above the water, the height of today's tallest skyscrapers. The exposed ice in this configuration does not have the strength to support itself. The ice shelf itself normally provides a buttress of support, but after it disappears, the exposed face of the glacier becomes structurally unsound and rapidly disintegrates, falling into the sea. This leaves a newly exposed and unstable ice face that also collapses. This process repeats. Because the Thwaites Glacier helps hold nearby glaciers in place, these glaciers eventually collapse, too, sparking a chain reaction. The collapse of the Thwaites Glacier can raise global sea levels 2 feet (61 cm) all by itself. The collapse of other glaciers exacerbates this sea-level rise.

The presence of warm water beneath the glacier only makes the situation worse. Circulation patterns in the ocean bring warmer water from the north to Antarctica, and these warmer waters can undermine and melt the ice at the front of its huge glaciers.

BELOW Glaciers in Antarctica flow downhill toward the ocean due to gravity. When they reach the ocean, an ice shelf extends out into the water. Warming ocean water has started undermining the glacial edge and accelerates glacial movement.

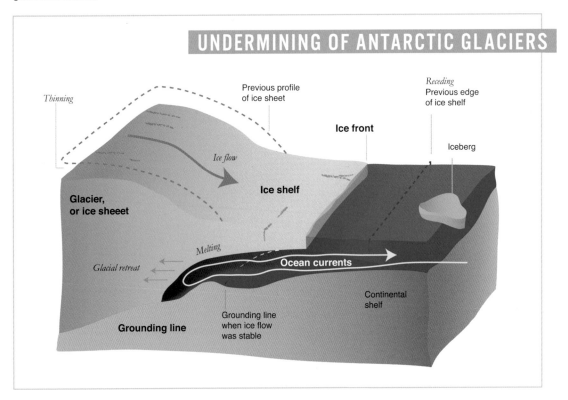

UNDERMINING OF ANTARCTIC GLACIERS

Thinning

Previous profile of ice sheet

Receding
Previous edge of ice shelf

Ice front

Iceberg

Ice flow

Ice shelf

Glacier, or ice sheeet

Melting

Glacial retreat

Ocean currents

Continental shelf

Grounding line

Grounding line when ice flow was stable

Scientists have also discovered sources of volcanic heat under Antarctica. Although their impacts are not yet fully understood, heat under the ice is not good, no matter what.

In January 2020, scientists discovered just how quickly ice is melting by drilling a hole through the Thwaites glacier and then lowering a robotic camera into the water below. This was the first time that scientists have been able to see underneath this enormous glacier and measure different parameters. They found that when relatively warm water circulates under the ice shelf, it increases the space between the glacier and the continental land below. This gap lets the glacier move more quickly, because there is less friction. Normally, the glacier grinds against the rising land—in this case, the edge of the basin; this land holds the glacier back. But as the glacier melts from underneath, it has less contact with the basin. Without this contact with land, the glacier can move into the water more rapidly (again, rapidly for a glacier). As the glacier comes into contact with more water, another gap forms, and the cycle repeats. This process speeds up the natural pace of the glacier's movement. In the last ten years, the pace of the glacier's flow has increased by about 5 percent. Scientists fear that this acceleration brings the timeline for the collapse of the glacier forward from centuries to decades.

We find more bad news when we look at the rate of iceberg detachments from the ice shelves that help protect the glaciers from rapid collapse. The ice shelves seem to be deteriorating more quickly than expected because of warming oceans. In 2017, for example, one of the largest icebergs ever seen broke off the Larsen C ice shelf. The size of Delaware, it contains 1 trillion tons of ice. The ice shelf for the Thwaites Glacier also seems to be deteriorating rapidly—much more rapidly than expected—because

VOLCANOES UNDER ANTARCTICA

As if climate change weren't worrisome enough for Antarctica, it turns out that under the ice sheet there lurks another threat—volcanoes. One hundred thirty-eight volcanoes have been discovered so far, ninety-one of them in 2017 alone.

Right now, all of these volcanoes appear to be dormant, eruption-wise, but they can produce heat under the ice that accelerates melting. And if one of them were to erupt in a significant way, it could have an effect similar to a nuclear bomb.

of the warm ocean water. Over the last quarter-century, the rate of melting in places like this glacier has tripled.

The increasing rate of meltwater in Antarctica, especially West Antarctica, makes the situation worse. Over the last twenty-five years, Antarctica has shed 6 quadrillion pounds of meltwater, and the rate of melting is accelerating. To put this in global terms, 6 quadrillion pounds of meltwater is the equivalent of 8 millimeters, about a third of an inch, of global sea-level rise. A complete collapse of West Antarctica generally will raise sea levels by 10 or 11 feet (3 or 3.6 m), and that doesn't account for the sea-level rise from the Greenland ice sheet or East Antarctica. Scientists are tracking Antarctic ice with a great deal of concern, because all the trends are in the wrong direction.

We have two doomsday scenarios at hand: the natural one, which unfolds more slowly over decades or centuries, as ice in Antarctica (and around the world) melts more rapidly due to climate change; and then the man-made version, where a nuclear bomb in the region of West Antarctica rapidly accelerates the timeline. Volcanic activity is also a wild card that may become important as we discover more about the volcanoes under Antarctica.

Today, some coastal cities are preparing for sea-level rise by constructing gigantic seawalls to hold the rising ocean back. New York City is considering a sea wall that would be 6 miles (9.6 km) long and cost $20 billion per mile to build. Similar plans are forming for cities like Staten Island, New Orleans, and Seattle. But these are stopgap measures. What if humanity were to prevent sea levels from rising altogether? In this case, more action needs to be taken.

For the scenario that unfolds naturally, humanity would need to eliminate the release of carbon dioxide *into* the atmosphere and simultaneously take the opposite action: start removing carbon dioxide *from* the atmosphere. As you'll learn in Runaway Global Warming (see page 52), carbon dioxide is the primary driver for the rising temperatures and warming oceans that threaten Antarctica. Perhaps humanity would need to take more extensive steps to try to slow climate change even more by turning to geoengineering (see page 63 for a complete discussion).

Solving the man-made scenario is potentially more difficult. Antarctica is an enormous place largely devoid of humans and without its own government. It measures 5.4 million square miles (14 million km²), nearly double the size of the continental United States, and has about 15,000 miles (24,150 km) of shorelines. At any given moment, there might be at most a few thousand people inhabiting this gigantic area. In the winter, the number falls into the hundreds.

A nuclear bomb could arrive by many different routes. Many nations now have ballistic and cruise missiles that can land nuclear bombs on Antarctica. Or a bomb could be dropped from an airplane. It is also easy to approach Antarctica by boat, especially in the summer. In 2018, Colin O'Brady made history by hiking 932 miles (1,500 km) across Antarctica in 54 days while dragging an enormous sled of equipment and supplies behind him. The trip took him from the Ronne Ice Shelf to the South Pole and the Ross Ice Shelf. Given O'Brady's feat, the possibility of transporting a bomb in on foot cannot be ignored.

It would be exceptionally difficult and expensive to protect all of Antarctica from all these threats. For example, Antarctica could be supplied with missile defense systems to counter a ballistic attack, but this system would be extremely expensive to build, and then an entirely separate system would be needed to defend against cruise missile attacks. With a radar system and air force, Antarctica could intercept a rogue airplane, and a fleet of drone boats could surround Antarctica with a line of deterrence against boats, but given the size of the continent, this too would be enormously expensive. These strategies would require Antarctica to establish a governing body and then fund an impressive military presence for self-defense. The cost of building a defense system big enough to defend a continent this large would be enormous. During the coldest parts of the

ABOVE It is extremely easy to drive a boat right up to a glacier in Antarctica, especially in the summer. If the boat contains a nuclear bomb that is able to destabilize a massive amount of ice, it could potentially raise sea levels across the planet.

year, even the most dedicated and disciplined military would be hard-pressed to keep systems operational. With the existence of highly destructive weapons such as Russia's tsunami-generating submarine (see page 157), it is unclear how Antarctica (or any coastal city) can be protected from these threats without an extensive global response. Therefore, diplomacy may be the only possibility.

This is the problem with, and the nature of, terrorism. It can take ten years and billions of dollars to build a skyscraper. But, as we have seen all too well by direct example, a terrorist cell can steal a passenger jet and completely destroy the skyscraper in a couple of hours, essentially at no cost. Despite how

unfair this is, the only solution is to anticipate the attacks and then spend the resources to deter them all. Either we track and closely monitor every single person on the planet with terrorist intent, or we defend every vulnerable target. Or both. It is a tremendous challenge.

And even if terrorists do not act, Antarctica is melting as we speak. The melting that we see today and the rate at which the melting is accelerating is the fault of human activities. And here humanity faces another important choice: it must take the steps necessary to lower global temperatures or accept the catastrophic losses when runaway melting causes global sea levels to rise.

EMP ATTACK

You are sitting in rush-hour traffic on a winter evening, driving home from work just like any other day. On this night, you are running a little late because of a last-minute emergency in the office. It is 6:30 p.m., and daylight is in short supply. All the streetlights and buildings downtown light up the city streets on an otherwise moonless, starless, deep-dark night. And then there is a bright flash in the sky.

Suddenly all of the lights in the city vanish, and your car stops working. The engine cuts off, the radio dies, the headlights go dark, and the dashboard extinguishes. You hit the brakes hard to avoid ramming the car in front of you. The sudden utter darkness makes it difficult to see. Seconds later, you lurch forward when the car behind you smashes into your rear bumper. You are rattled again when another car bumps into the one that just hit you.

The scene outside your window is surprising. All of the cars have come to a standstill and gone dark. All of the nearby buildings and streetlights are dark. It is nearly pitch-back on this city street. The only illumination comes from a circle of light from one lone car ahead in the distance.

You reach for your phone to use the flashlight, but like your car, your phone is completely dead. The screen is black and unresponsive. You press the power button, but nothing happens.

You open your car door and stand up. Other drivers are doing the same, and everyone is bewildered. Now that you are standing, you can see that it is not just this street that has been affected. The entire city, at least the wide swath of it that you can see, has gone completely dark.

You hear a huge crash behind you and the sound of glass breaking. A gunshot and then another. Someone screams, and people start shouting. You duck back into your car, slam the door, and instinctively hit the lock button to lock the doors. Nothing happens. You reach forward and feel for the manual door lock, then find the one on the passenger side too.

As you sit there, you realize that this is no ordinary blackout. In a typical power outage, cars are fine. Cell phones are fine. But now, nothing is working. The thousands of cars around you are permanently gridlocked. Your phone is dead, too. Because this isn't the best part of town, the looting has already started. People are lighting things on fire for light, and they are breaking store windows to start stealing the merchandise. The speed of the collapse is impressive and also terrifying. You hear several gunshots, and more glass breaking. Now what?

RIGHT A scud missile like this one on its mobile launcher is 37 feet (11.3 m) long and has a range of 200 miles (322 km) or more, depending on the model. It can carry a one-ton payload, enough for an EMP device that can impact a city or small country.

An EMP attack—or an electromagnetic pulse attack—is widely believed to be a doomsday scenario for a developed country like the United States. In theory, a large-scale EMP attack could bring an entire continent to its knees. If you have read the best-selling book *One Second After* by William R. Forstchen or seen the movie *E.M.P. 333 Days,* you are familiar with the common depiction of an EMP attack on the modern United States. The description on the previous page presents the way that many have imagined the immediate aftermath of this doomsday scenario would go.

Although they sometimes exaggerate certain details for dramatic effect, these depictions have a basis in fact. Military and civil defense leaders see EMP attacks as an issue to take seriously, and in the United States they were working to harden the nation against the threat. This scenario even made it into the Republican Party's platform in the 2016 election. It describes an EMP attack as "no longer a theoretical concern—it is a real threat."

The most commonly proposed mechanism for implementing an EMP attack is the detonation of a nuclear bomb anywhere between 10 and 300 miles (16 and 483 km) above the ground, with the height of the explosion determining how widespread the effects are. EMP effects are line-of-sight, so the higher the detonation, the larger the impacted area.

When a nuclear bomb goes off high in the atmosphere or even out of the atmosphere in low Earth orbit, it creates a powerful pulse of electromagnetic energy that has the potential to take out every electrical device in our complex technological society. The immediate effect is a power blackout. The long wires used to transmit and distribute electricity act as antennas, picking up and transmitting electromagnetic energy to everything connected to the power grid, from motors and refrigerators to handheld electronics and computers. This can cause widespread failures in the grid itself as well. The same happens for appliances and devices that aren't plugged into a socket: smartphones, laptops, GPS systems, cameras—they stop working too. The pulse can send the delicate transistors in microprocessor chips into overdrive and short them out.

Unfortunately, this blackout is long-lasting and would affect most of our modern devices. In 2017, the United States House of Representatives commissioned a report titled "Empty Threat or Serious Danger: Assessing North Korea's Risk to the Homeland." The report contains sobering predictions from experts about the effects of

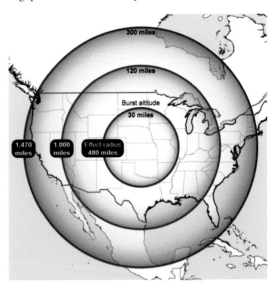

ABOVE This map shows areas that would be affected by an EMP from a nuclear bomb. The higher the altitude of the burst, the larger the affected area.

EMP attacks, along with recommendations for the future. For example, they predict: "The result could be to shut down the US electric power grid for an indefinite period, leading to the death within a year of up to 90 percent of all Americans—as the EMP Commission testified over eight years ago." In a statement from 2015, Dr. Peter Vincent Pry says: "The EMP Commission estimates that a nation-wide blackout lasting one year could kill up to 9 of 10 Americans through starvation, disease, and societal collapse."

These are worst-case scenarios, but they are plausible in the event of long-term power disruptions and the resulting societal breakdown. For example, in terms of food, the second report states: "Worst of all, about 72 hours after the commencement of blackout, when emergency generators at the big regional food warehouses cease to operate, the nation's food supply will begin to spoil. Supermarkets are resupplied by these large regional food warehouses that are, in effect, the national larder, collectively having enough food to sustain the lives of 310 million Americans for about one month,

at normal rates of consumption." If the food delivery network shuts down due to lack of power and lack of trucks, supplies dry up in weeks. Without a functioning power grid, city water systems stop working. The supply of medicines is cut off like food is, and without electricity, hospitals cannot do anything but the simplest procedures. The lack of access to ATMs and other financial transactions, such as credit card payments, also denies people access to money.

One item that would still work after an EMP attack is guns, and guns may be the biggest worry in any event that takes down the power grid. Guns are purely mechanical devices. They use an old-school chemical propellant in the form of gunpowder, which has been around for centuries. Therefore, guns are completely immune to an EMP attack. There are approximately 400 million guns in the hands of American citizens—more guns than people. A long-term power outage can lead to unrest (see Grid Attack, page 90); the fear is that anarchy and looting would take over because of the hunger and desperation that can occur in a long-term power failure.

SCIENCE

How can a nuclear bomb cause catastrophic electromagnetic effects in the first place? What is different about a nuclear bomb that makes it produce an EMP effect?

The answer is that, unlike other explosives, nuclear bombs release a massive amount of gamma radiation at the time of the explosion.

We are all familiar with the electromagnetic effect known as visible light. This is what our eyes and cameras can detect. Atoms under certain conditions

emit photons in the visible spectrum. For example, if you heat up a piece of iron to a high enough temperature, it will emit a red glow and then, if hot enough, a white glow. The heat causes photons to be released.

To the right of visible light on the electromagnetic spectrum is ultraviolet radiation. Although invisible to the human eye, UV light is strong enough to damage DNA and kill skin cells, causing sunburn. If we move farther along the spectrum, we

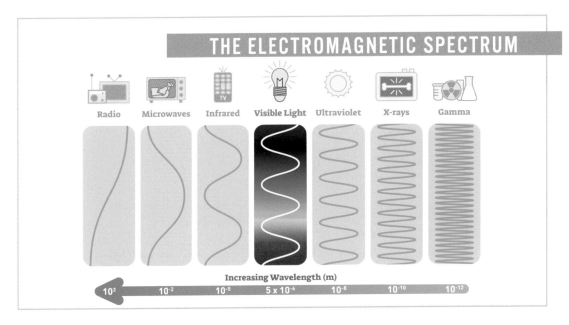

THE ELECTROMAGNETIC SPECTRUM

Radio   Microwaves   Infrared   **Visible Light**   Ultraviolet   X-rays   Gamma

Increasing Wavelength (m)

10^3 10^{-2} 10^{-5} 5×10^{-6} 10^{-8} 10^{-10} 10^{-12}

ABOVE Waves on the left side of the electromagnetic spectrum have longer wavelengths, whereas waves on the right side of the electromagnetic spectrum have shorter wavelengths, giving them the ability to pass through objects and ionize atoms.

have X-rays. X-rays are strong enough to penetrate and pass through many objects, ripping electrons away from atoms in the process.

X-rays are an example of ionizing radiation; an atom that is missing electrons is an ion, hence the name. The missing electrons will break chemical bonds and create free radicals. In the human body, ionizing radiation leads to DNA damage and mutations, as well as cell death. This is why X-rays are detrimental to humans in larger doses and why you wear a lead apron when you receive X-rays at the dentist. The lead decreases your X-ray exposure, as the density of lead prevents the radiation from penetrating your body.

At the far right of the electromagnetic spectrum are gamma rays, another form of ionizing radiation that is even stronger than X-rays. X-rays can't pass through a thin 2-millimeter sheet of lead, while gamma rays will happily fly right through. It would take a sheet of lead with a thickness of more than a foot (300 mm) to shield against gamma radiation.

An X-ray photon has 100 kilo-electronvolts (KeV) of energy or less; a gamma ray can be at least ten times more energetic, typically with 1 to 3 mega-electronvolts (MeV) of energy. Gamma rays emitted from objects like exploding stars are even more energetic.

When a nuclear bomb detonates, the splitting of atoms during the explosion creates gamma rays (see page 29). This gamma radiation interacts with atoms in the atmosphere. It strips off their electrons, and these electrons shoot off with high energy. This stream of electrons forms an electric current. Normally, we think of electric current as electrons moving in a wire, and this movement creates electromagnetic waves (this is how antennas work, for example). The stream of electrons moving in the air also creates electromagnetic waves. Since the gamma rays release incredibly energetic electrons, and many of them in an instant, the flow of these high-energy electrons in the atmosphere creates an enormous electromagnetic pulse.

This pulse arrives on the ground in three phases:

- **E1:** During the E1 phase, a brief high-power pulse occurs. Think of this as an incredibly powerful lighting strike, but one that hits everything at once. If you have a device like a computer plugged into an outlet with lightning or surge protection, this pulse is so powerful that it would probably destroy and bypass the protection you have.

- **E2:** This pulse is very much like the pulse emitted after a normal lightning strike—and, like E1, it hits everything at once. Surge protectors would be ineffective against it, as they were likely to have been blown out in E1. The difference between E1 and E2 is a matter of degree, with E1 being much more intense. Though E2 is less powerful, it can still be catastrophic in its own right.

- **E3:** The blast distorts the Earth's magnetic field. The field then recovers to its normal state, creating electrical currents in conductors as the distortion releases. These currents accumulate in any long wire (such as power lines), destroying everything connected to the wire.

The fact that an EMP attack has this three-pronged effect is what makes it so dangerous and devastating. Not only does it affect the power grid and everything connected to the power grid, it can also impact unshielded devices that are not connected to the grid at all.

A nuclear bomb and its gamma rays represent one way to create an electromagnetic pulse across a wide area. But if you are on a budget and want to affect only a city rather than a continent, an alternative is a flux compression generator. This device combines a large coil of wire, a capacitor bank to instantly charge the coil, and a conventional bomb to destroy the coil. (There are several other architectures as well.) The coil generates a strong electromagnetic field by itself. The bomb compresses and destroys the coil in such a way that this field is magnified enough to emit an electromagnetic pulse that has a range of several miles.

BELOW An EMP would affect any unshielded components on the power grid, including this equipment at a high-voltage electric substation.

One interesting thing about the EMP scenario is that we are not exactly sure what would happen if a full-scale EMP attack occurred today. The United States and Russia detonated nuclear bombs in the upper atmosphere and in space in the 1950s and 1960s. Most notably, the United States detonated a nuclear warhead 250 miles (402 km) above the Pacific Ocean on July 9, 1962—an experiment known as Starfish Prime. This warhead had a yield equivalent to 1.4 megatons of TNT, which is roughly 100 times more powerful than the bomb used at Hiroshima.

The Starfish Prime bomb produced an EMP effect. In Hawaii, nine hundred miles (1,450 km) away, three hundred streetlights burned out and a telephone link to the mainland was put out of commission. But Hawaii's power grid did not collapse. And, though many Navy ships were stationed under the bomb when it went off, they were not disabled. Meanwhile, the Soviet Union saw different results when it conducted similar tests as part of Project K, using smaller, 300-kiloton bombs and also detonating them in space. One test on October 22, 1962, destroyed a power plant, telephone lines, and power lines.

Keep in mind that in 1962, the earliest electronic chips had just been invented and microprocessors did not exist yet. Computers were extremely rare rather than ubiquitous. Today, electronic devices form the foundation of our daily lives, making us more susceptible to the effects of an EMP. Microprocessors now control everything from car engines to smartphones to factories. These intricate devices are fragile: even a spark of static electricity can destroy a microprocessor. How would they fare during an EMP attack?

In the "Report of the Commission to Assess the Threat to the United States from Electromagnetic Pulse (EMP) Attack," professionals in the United States ran experiments to assess various vulnerabilities. The good news is that things seem to be improving. Today's microprocessors are well-shielded, mainly to prevent them from spreading interference to other electronic devices, and this shielding can also protect against EMPs.

What about cars? Thanks to the 1963 Partial Nuclear Test Ban Treaty, no country conducts nuclear tests like Starfish Prime anymore, but the US military has carried out research on the effects of an EMP strike. Tests with modern automobiles subjected to EMP fields seem to indicate that many cars would be unaffected, especially if they are turned off when the pulse arrives. In one test, just three of the thirty-seven cars studied stopped running, and only if the fields were powerful enough. The researchers noted: "In an actual EMP exposure, these vehicles would glide to a stop and require the driver to restart them." Most cars "exhibited malfunctions that could

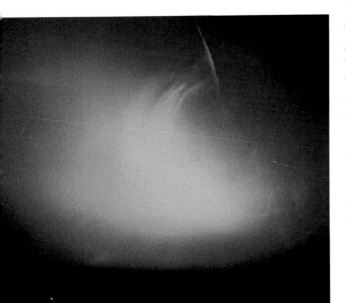

LEFT The blast from the Starfish Prime bomb damaged streetlights in Hawaii and caused telephone outages.

be considered only a nuisance (e.g., blinking dashboard lights) and did not require driver intervention to correct."

Modern telephone networks might also withstand the worst of an EMP attack. The report explains that "damage to telephones, cell phones, and other communications devices would not be sufficient to curtail higher than normal call volumes on the civilian telecommunications network after exposure to either low or high E1 EMP levels." The problem described here is a social one rather than a technological one. As soon as there is an EMP attack, millions of people will start calling one another, along with various emergency services. It is the volume of these calls that would overwhelm the system, rather than the system itself breaking down. The report's authors predict that "the remaining operational network would be subjected to high levels of call attempts for some period of time after the attack, leading to degraded telecommunications services." There has been little public guidance about how citizens need to behave in the event of an EMP attack, but at the very least we need to train ourselves as a society to not immediately jump on the phone in the event of an attack. In addition, it would be prudent to build extra capacity into the network to handle a huge emergency.

What about the power grid? As you'll learn in Grid Attack (page 90), the United States has a power grid that is huge and multi-dimensional, and most of it is outdoors. The interconnected nature of these components creates a "weakest-link" problem. If one important part of the power grid fails, the whole system can be taken out. As with the Fort Lake Peck scenario (see page 9), there is also the problem of a cascading failure. If one major transmission line fails, its load shifts to the remaining lines. This in turn could overload these lines, causing them to stop working and creating a chain of failures that incapacitates the whole grid. This scenario has played out more than once in the United States, most famously in the 2003 Northeast power blackout. When this has happened in the past, it usually took a few days to recover. In an EMP attack, with multiple failures occurring simultaneously over a large area, it might take months.

Some countries have begun considering how they can prepare their power grids for an EMP attack. The Republican Party's 2016 platform observes how "hundreds of electrical utilities in the United States have not acted to protect themselves from EMP . . . . The President, the Congress, the Department of Homeland Security, the Department of Defense, the States, the utilities, and the private sector should work together on an urgent basis to . . . protect the national grid and encourage states to take the initiative to protect their own grids expeditiously." Individual states in the United States are stepping up, as they act as the primary regulators of power companies. For example, state-level regulators can require power companies to add the equipment necessary to protect large transformers from EMP and coronal mass ejection (CME) events (see page 158). Another approach to consider is shielding any backup power systems. At the federal level, the president issued an executive order to increase the country's resilience to EMPs in 2019. The UK has also recognized the threat of an EMP attack from terrorists or a rogue state.

In the ideal case, where would we be? The United States, along with governments across the world, would have power grids, telephone networks, automobiles, and electronic devices that are all immune to the effects of an EMP attack. When we arrive at this point, the doomsday scenario attached to an EMP attack disappears.

RUNAWAY GLOBAL WARMING

THREAT LEVEL: WORLD

As a teenager living in 2070, I cannot feel anything but disgust for our grandparents and great-grandparents—the people who oversaw the destruction of Planet Earth rather than acting to save it. When they were younger, they were living in a global paradise compared to today. The planet that we have inherited from them is an unmitigated disaster, completely different—and completely appalling—compared to the planet fifty years ago, never mind what the planet was like before the twentieth century.

In history class, watching archival videos and images is unbelievably heartbreaking. People in the 2000s and 2010s could casually walk through cities like New York City or San Francisco. It is amazing that such gigantic cities could exist out in the open. But the thing that is so unsettling is the profligate waste and destruction that is so clearly on display. People actually owned private automobiles and burned gasoline in them! There are home videos that show families living in their own private houses, where they could walk out into their private backyards and barbeque a meal while birds chirped, bugs flew by, and squirrels played in the trees. Citizens of the 2010 era could also walk through national parks of startling beauty and diversity. There are clips of Glacier National Park, the Amazon rainforest, the ice sheets in Greenland and Antarctica. The thing that gets me most are the videos of divers in the ocean interacting with sea life. These nature videos show stunning beauty as well. There were whales—actual whales!—along with coral and schools of fish in the ocean, before the great die-offs. Herds of elephants, rhinos, giraffes, and zebras roamed free on the Serengeti before they all went extinct. It is unimaginable now that these cities, homes, parks, and animals actually existed. This was a paradise that no one today can imagine, now that Planet Earth has been destroyed.

Do we have a life today? Are we able to live? Of course. Humans are survivors. But we live on an overheated, sterile planet today compared to the diversity of life seen in centuries past. Through human stupidity, greed, and infighting, we have watched paradise slip through our fingers. Most of the wildlife on the planet is gone. The rainforests are baked deserts. There are large areas of the planet where you must wear a suit or drive in a vehicle when outside because the heat makes air-conditioning a necessity for survival. Now we, as teenagers, inherit a wasteland.

Warning sign after warning sign was ignored until it was far too late to stop the process of runaway warming. Once the runaway part got under way, there was nothing that could stop it, and the

RIGHT Emissions from a coal-fired power plant billow into the sky. A typical 1-gigawatt plant emits approximately 6 megatons of carbon dioxide per year.

planet was lost because people could not muster the political will to solve the problems. Today, the remnants of the human species all live precariously, avoiding the scorched or flooded remains of the planet's surface. Now we, the survivors, look back in anguish, at the many lost opportunities that were frittered away, wishing for the paradise that once was Planet Earth.

SCENARIO

Scientists have been talking about global warming and global climate change since the 1970s. The term *global warming* was coined in 1975 by geochemist Wallace Broecker. It refers to the upward trend in average global temperatures. Then in 1979, meteorologist Jule Charney adopted the term *climate change* to encompass other effects: for example, a change in rainfall that makes flooding or drought more common, or a greater incidence or strength of hurricanes. However, the study of what we now call global warming goes back quite a bit further. The original recognition that adding carbon dioxide to the atmosphere might cause warming can be attributed to Eunice Foote, who wrote about the heat-trapping effects of CO_2 in 1856.

In 2006, Al Gore, former vice president of the United States, chose to champion environmental activism and broaden public awareness of climate change with his documentary *An Inconvenient Truth*, his accompanying book of the same name, and a speaking tour. For many people in the general public, this was their first real encounter with this problem, the data supporting it, climate science in

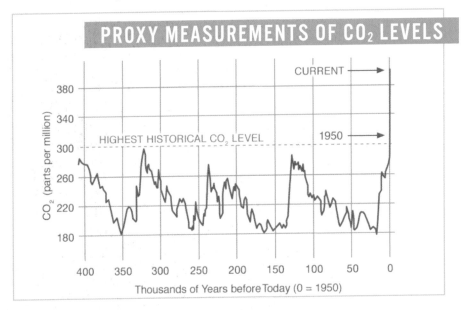

ABOVE Using data gathered from ice cores, this graph shows the extent to which atmospheric carbon dioxide levels have grown over the last century.

HOW DO SCIENTISTS MEASURE CO₂ IN THE ATMOSPHERE?

To accurately measure the level of CO_2 in the atmosphere, we first need to gather a good sample of "air." We cannot go down to the street corner to gather the sample, because any city location will have compounds from factory emissions, human emissions, and automobile emissions that would skew the result.

NOAA's Earth System Research Laboratory therefore takes a sample of air from a mountaintop in Hawaii that is 11,000 feet (3,350 m) above sea level. All moisture is removed from the sample, and the dry air is placed in a test cell that has a very precise infrared light source on one side and a very precise infrared detector on the other. Since CO_2 absorbs infrared light, scientists can look at the readings from the infrared detector to determine the concentration of CO_2 in that air

sample. (The moisture is removed from the sample because moisture also has a greenhouse effect and therefore also absorbs infrared light.)

ABOVE The NOAA Earth System Research Laboratory at Mauna Loa uses the advantages of its site, including its remote location and altitude, to monitor greenhouse gases.

general, and the predictions of impending doom. General public awareness has been growing since then, aided by unavoidable evidence that causes people to take notice. It is easy to see the patterns of melting ice in the Arctic, for example, or the fires in the rainforest, and a graph of global average temperatures shows undeniable trends.

In a nutshell, this is how global warming works:

1 **Human activities add carbon dioxide (CO_2) to the atmosphere.** For example, the world pumps approximately 100 million barrels of oil out of the ground every day, refines much of it into fuel, and then burns it. This process adds approximately 40 gigatons of carbon dioxide to the atmosphere per year. There are other sources of carbon emissions as well, including

coal, natural gas, and the burning of forests like the Amazon rainforest (see page 178), which releases carbon sequestered in the living trees.

2 **The amount of carbon dioxide in the atmosphere accumulates over time.** The level of carbon dioxide in Earth's atmosphere was about 280 parts per million (ppm) in the year 1800. By the early 1900s it had risen to 300 ppm. Four hundred ppm was considered by the scientific community to be a "red line," and humanity crossed it sometime in 2015–2016. Above this level of CO_2, any hope of keeping global temperature rise within a "reasonable level," defined by the Paris Agreement to be a rise of 2°C (3.6°F) or less, becomes impossible.

3 **The CO_2 in the atmosphere is a greenhouse gas.** In a glass greenhouse, sunlight (visible light) comes in through the glass and converts into heat (infrared light). This heat cannot escape because of the glass—glass is less transparent to infrared light than it is to visible light. This causes the greenhouse to heat up. Carbon dioxide has a similar effect in the atmosphere, letting visible sunlight through but keeping heat in. Increasing concentrations of CO_2 in the atmosphere intensify the planet's greenhouse effect, causing global average temperatures to rise. Average global temperatures started rising noticeably around 1980 after a steady period that stretched from 1940 to 1980. In 2016, global average temperatures were about 1°C (3.4°F) higher than in 1980.

Runaway global warming occurs when we have reached a tipping point in carbon emissions. We reach this tipping point not only because humans release increasing amounts of CO_2 that raise temperatures all by themselves but also because the rising temperatures engage positive feedback loops that cause global warming to accelerate. There are five feedback loops that scientists discuss most often:

1 **Once warming temperatures start melting permafrost, large amounts of sequestered carbon become available.** Bacteria turn this carbon into CO_2 or methane, a greenhouse gas twenty times more powerful than carbon dioxide. Therefore, thawing permafrost has the potential to release large amounts of greenhouse gases. As megatons of CO_2 and methane start entering the atmosphere from melting permafrost, the warmer conditions on Earth cause more permafrost to melt more quickly, increasing greenhouse gases and temperatures further. The melting permafrost thus causes more permafrost to melt in a positive feedback loop.

2 **As ice in the Arctic and Antarctic regions melts, less of the Earth's surface is covered by the white color of the ice.** White ice reflects a lot of sunlight back into space, whereas brown dirt or blue ocean water absorbs more light, creating heat. As ice melts, the planet warms more quickly, which in turn causes more ice to disappear, and so on.

3 **For the past several decades, oceans have been absorbing carbon dioxide.** Though this means there is less CO_2 in the atmosphere, the oceans are becoming more and more saturated with carbon dioxide and the rate of absorption is slowing. As this occurs, the CO_2 that humans release into the atmosphere remains in the air, accelerating the rate of

heating on the planet. This accelerates the melting of permafrost and ice, exacerbating the previous two feedback loops.

4 **Temperatures will reach a point where they cause forests to start dying, from either heat or drought.** As forests die, and especially as dead forests burn, they release their sequestered CO_2 into the atmosphere, which causes more warming, which makes forests even more susceptible to collapse.

5 **Clouds also reflect sunlight into space, but as temperatures rise, the processes that form clouds change, and clouds start to disperse.** As clouds disperse, more sunlight hits the Earth, which causes temperatures to rise. This then causes even more clouds to disperse, and so on.

At one level, the solution to global warming seems simple enough: If temperatures are rising because humans are pumping greenhouse gases like carbon dioxide into the atmosphere, we can stop pumping CO_2 into the atmosphere and take some CO_2 back *out* of the atmosphere to fix the problems it causes.

But once these positive feedback loops get under way, it may not matter if humans curtail the use of fossil fuels, or if humans were to start planting trees. The amount of carbon locked in permafrost, for example, exceeds all of the carbon currently in the atmosphere. Therefore, once permafrost starts melting at a sufficient rate, the melting permafrost alone can continue adding greenhouse gases to the atmosphere all by itself. Global warming could become unstoppable.

In the worst-case scenario, with all of these feedback loops engaged, melting ice causes sea levels to rise to a devastating degree. The Greenland ice sheet alone would raise sea levels by 24 feet (7.2 m) once it fully melts. The West Antarctic ice sheet and East Antarctic ice sheet would increase them by another 16 feet (4.8 m) and 175 feet (53 meters), respectively. And this does not factor in other glaciers around the world. If everything melts, sea levels could increase by 215 feet (65 m). All of Florida disappears underwater. Major cities like New York City, Washington, DC, Shanghai, London, and Mumbai are all lost. Tens of millions of people in coastal areas around the world lose their homes and land and must relocate. Rising seas can ultimately dislocate half a billion people.

Twenty years ago, scientists speculated that a meltdown like this would take millennia. But the melting has accelerated rather than holding steady.

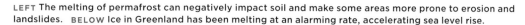

LEFT The melting of permafrost can negatively impact soil and make some areas more prone to erosion and landslides. BELOW Ice in Greenland has been melting at an alarming rate, accelerating sea level rise.

THE WORST WORST-CASE SCENARIO

Imagine that global average temperatures continue rising, well past the "safe" level of 2°C (3.6°F). If ocean surface temperatures rise high enough, then a moisture feedback loop could arise. Moisture in the atmosphere, it turns out, is a very efficient greenhouse gas. As air temperatures rise, the air can hold more moisture. As ocean temperatures rise, more water evaporates, causing more of a greenhouse effect. Eventually the temperatures rise high enough to where a runaway greenhouse effect occurs, with the evaporating oceans feeding into an upward temperature spiral. Hopefully, humans will take action long before things get this far.

For example, between 1980 and 1990, 51 billion tons of ice from Greenland's ice sheets melted into the ocean, but between 2010 and 2018, this figure increased to 286 billion tons. With ice melting at an accelerated rate and intensifying the positive feedback loops described earlier, sea levels end up rising much faster than previously predicted.

Melting ice and rapidly rising sea levels are just one aspect of the worst-case scenario. Rapidly rising temperatures may also make large areas of Earth uninhabitable to human beings, animals, and plants.

There is a narrow range of temperatures where life can thrive; once we exceed these temperatures, life on the Earth's surface starts shutting down. Imagine walking through a humid jungle. Because you are walking, you are using your muscles, and they generate heat. Your body tries to cool off by sweating. But since it is humid already, your sweat does not evaporate much. If the outdoor temperature is lower than your body's, your body can manage the heat. When the temperature climbs higher, things become iffy. If the temperature reaches 95°F (35°C) or higher, human activity in a humid environment becomes impossible without overheating. Human beings simply cannot operate for any length of time if it gets this hot in a very humid place. And neither can other animals and, eventually, plants. The entire equatorial region of the planet could potentially become uninhabitable. Warm, humid places like India, Bangladesh, and Pakistan are significantly vulnerable, as are the Amazon rainforest, big parts of equatorial Africa, and even the southern United States. The Middle East is not humid, but the temperatures in some Middle Eastern countries, such as Qatar and Saudi Arabia, are already extreme, and they will only get worse as global average temperatures rise. We may come to a point, in the not-too-distant future, where humans have to wear special suits to provide cooling so they can walk around on certain parts of Earth on hot days. It simply would not be possible for humans to survive outside without an air-conditioned suit or vehicle.

In addition, global warming will amplify other doomsday scenarios described in this book. Rising carbon dioxide levels will create more intense hurricanes (see page 190), kill off tropical forests located in and around the equator (see page 178), acidify the oceans (see page 200), and cause the Gulf Stream to collapse (see page 208). Its far-reaching impact makes it a huge problem with dire consequences, unless humanity takes action in time.

You may have heard about a lack of consensus, or a difference of opinion, about whether global warming and climate change are happening. On one side of this debate is a large majority of the scientific community, which has a great deal of evidence and data to back up their concerns. On the other side of the debate is a wide-ranging group that includes laypeople who are not aware of the evidence, radio personalities, as well as a number of large corporations, especially in the fossil fuel industry, that benefit from the status quo and use their money to maintain it.

So, why are scientists so sure? What leads them to the undeniable conclusion that global warming and climate change are real? It can be helpful to look at the interlocking pieces of evidence that the scientists have observed.

The first piece of evidence is the rising amount of carbon dioxide in the atmosphere. The concentration of CO_2 is easy to measure, and it has been rising year after year. In addition, there is no question that carbon dioxide in the atmosphere acts as a greenhouse gas, trapping heat rather than allowing it to escape. This phenomenon has been studied for more than 150 years.

BELOW A severe drought and heat wave in 2015 devastated farmland in the state of Maharashtra in India. Temperatures rose as high as 104°F (40°C). As global warming worsens, seasonal temperatures in regions with subtropical climates are expected to exceed this level regularly.

We can find additional evidence in our current weather patterns and an undeniable upward trend in global average temperatures for the past forty years. There have been a number of record-setting heat days. In many places around the world, high temperatures are setting new records that are then broken again and again, even in Antarctica.

There has been an increase in extreme weather events. (see Hurricanes and Typhoons, page 190). A heavy downpour today is heavier than it was fifty years ago, and the number of heavy downpours is on the rise as warmer weather dissolves more moisture into the atmosphere, a process that leads to more rain. Even the lengths of the seasons are changing. In Australia, for example, scientists can compare the length of the summer in the 1950s with the length of the summer in the 2010s. Summers are now a month longer, and winters are a corresponding amount shorter.

As we discussed earlier, evidence of global warming can be seen in the melting ice sheets in Antarctica and Greenland, and the increasing rate of melting. Greenland's loss of ice was in the range of 100 gigatons of meltwater per year in the 1990s; it is now closer to 300 gigatons. Antarctica is experiencing a similar ice loss as summer temperatures reach new extremes. Scientists have observed that permafrost is melting at unprecedented rates as well. Permafrost got its name because it was thought to be permanent. But now, global warming has caused large areas of permafrost to melt. As noted earlier in the chapter, one big concern with melting permafrost is the positive feedback loop that it can create.

One side effect of the melting ice sheets is rising sea levels, which are currently increasing about 3 millimeters (0.1 inch) per year. Sea level rise is already affecting places like Norfolk, Virginia; Venice, Italy; and Miami, Florida.

Evidence of global warming can be seen in the oceans as well. Average ocean temperatures are rising alongside the atmospheric temperatures. Some of the excess carbon dioxide in the atmosphere gets dissolved in the ocean, causing ocean acidification (see page 200). Both of these trends align with what scientists observe in the atmosphere, and independently gathered data on these marine trends support the findings from atmospheric data.

Finally, farmers are noticing changes in growing areas all over the world. Wisconsin, for example, is the number-one area in the world for growing cranberries. But the winters grow warmer, there are more frequent hailstorms, and intense rainfall can cause flooding, all causing the environment in Wisconsin to deteriorate for cranberry growers. In a similar way, it is estimated that by 2050, half the land that grows coffee today will no longer support coffee plants. Potato farmers are seeing similarly worrisome effects as changes in rainfall and temperatures affect their crops. As noted above, important agricultural areas in Asia are on track to become so warm that humans will no longer be able to work the land as temperatures climb.

These pieces of evidence, and many others, show scientists that climate change and global warming are occurring on Planet Earth today and that human activities like fossil fuel consumption are the cause. A common rebuttal to this evidence is the idea that we don't know that humans are the cause of global warming. But we know, with certainty, that increasing levels of carbon dioxide in the atmosphere lead to warming. We also know with certainty that carbon dioxide is a greenhouse gas—that it leads to heat retention and therefore warming. It is easy to perform experiments and prove it. We also know, with certainty, that the amount of carbon dioxide in the atmosphere has grown from a preindustrial level of 280 ppm to 410 ppm today, a rise of 46 percent.

ABOVE Ice calves from Svalbard, an archipelago located in the Arctic Ocean.

This increase in carbon dioxide corresponds with and explains the sudden (in geological terms) rise in global temperatures. Someone might ask: "What if the sun is getting brighter? That would cause global temperatures to rise." But we know in several important ways that the sun is not getting brighter. These rebuttals fail in the face of insurmountable scientific evidence that points to human-caused global warming, with fossil-fuel consumption being the primary driver.

PREVENTION

The Paris Agreement in 2016 was an important milestone. It sought to limit global temperature increases to 2°C above pre-industrial levels by curtailing human emissions of CO_2. This was a good idea, and nearly every nation in the world signed on. But there is a problem: Having global average temperatures rise in lockstep with human activities is one thing; having them rise even faster because multiple positive feedback loops are in play is something else entirely. Given that humans are about to find themselves in this latter position, is there anything else that science can offer? What might humanity do to prevent or forestall the doomsday scenario where global average temperatures rise far more than anticipated or desired?

If we ask scientists these questions today, their advice would include several approaches:

1 Stop humans from destroying ecosystems that sequester carbon, such as the Amazon rainforest. Put an end to clear-cutting and burning forests.

2 Eliminate the use of all fossil fuels immediately.

3 Start pulling carbon dioxide out of the atmosphere and the ocean immediately.

4 Bring carbon dioxide levels back to pre-industrial norms (300 ppm or less).

In Rainforest Collapse (see page 178), we'll discuss how humanity can save carbon-sequestering rainforests from imminent destruction. The chapter on ocean acidification (see page 200) discusses ways we can stop using fossil fuels. In this section, we discuss

BELOW In this reforestation project, pipes act as browse guards to keep animals away from the seedlings.

the idea of extracting CO_2 from the atmosphere. These three approaches, if done quickly and with an accelerated effort, would accomplish the fourth item in a finite amount of time.

But what if humanity cannot do this? What if we cannot muster the political will or make the investments, and we stay on track for catastrophe? Or what if one or more of the positive feedback loops described above has already engaged irreversibly and we need to take more drastic actions to counter it?

In this case, humanity could turn to a set of tools broadly known as geoengineering. Geoengineering tries to control the climate through actions that go beyond the ordinary. Geoengineering takes direct steps to intervene on the environment—it is an active effort to try to lower the planet's temperature. For example, we might try to decrease Earth's global average temperature 2°C very quickly.

Here are six geoengineering possibilities that humanity might consider if the situation with global warming becomes more and more desperate:

ABOVE A view of Climework's direct-air capture plant in Switzerland. The plant removes carbon dioxide from the air and uses the captured carbon dioxide in a commercial greenhouse nearby.

1 If we look at volcanic eruptions (see page 132), we see that a big eruption releases sulfur dioxide and hydrogen sulfide into the stratosphere, where it can linger for a year or more. These gases, in sufficient quantities, reflect sunlight back into space and can cause a multiyear cooling effect. We could take advantage of this process by injecting a large amount of hydrogen sulfide or sulfur dioxide into the stratosphere ourselves, thus cooling down the planet almost immediately.

2 In a similar vein, humans could deploy large orbiting mirrors or swarms of smaller mirrors to reflect sunlight away from the planet. The advantage of this approach might be more control over global temperatures. Ground

stations could potentially instruct the satellites to angle their mirrors or even to leave orbit, if necessary.

3 Cloud brightening is a technique for increasing the amount of cloud cover over the ocean. The technique converts saltwater from the ocean into a fine mist that is released over specific areas of the ocean that need cooling: for example, a coral reef, or the Gulf of Mexico, or the hurricane-spawning portion of the Atlantic Ocean.

4 Ocean fertilization involves adding chemicals to the ocean to encourage the growth of plankton. The extra plankton would capture carbon and sequester when they subsequently

die and sink to the ocean floor. One such fertilizer is iron, which is generally scarce in the ocean but needed by plankton. Sprinkling iron dust in the ocean would cause plankton blooms.

5 Reforestation involves planting billions of new trees, which extract carbon dioxide from the air in the process of creating wood. In one proposal, humans could plant a trillion trees all over the planet to capture hundreds of gigatons of CO_2 from the atmosphere.

6 It is possible to create and deploy machines that extract CO_2 from the atmosphere and sequester this CO_2 or carbon. The technology is known as direct air capture. The advantage of these machines is that they take up far less land than forests do to capture the carbon. It is easy to imagine large solar or wind farms providing electricity in remote areas to power the machines that are capturing gigatons of carbon back out of the atmosphere.

Are these geoengineering approaches a good—or a viable—idea? With the exception of planting trees, all of these ideas will require research and effort to deploy. And we do not necessarily understand the unintended consequences and side effects. However, humanity may be forced to move forward with geoengineering.

The climate crisis that humanity has created is utterly enormous, and we need an equally enormous effort to solve the problem. Think about the Apollo missions to the moon during the 1960s and 1970s. At the program's peak, NASA had 400,000 people working to send humans to the moon. It was a monumental endeavor, with scientists and engineers inventing, testing, and deploying new technologies at an amazing pace. In just ten years, NASA invented from scratch the gigantic Saturn V rocket, all of its engines and fuel systems, the lunar lander, the space suits used on the moon walks, all of the life-support systems, and more.

Prior to the launch of the Sputnik satellite in 1957, humans had never sent anything into space. No human had ever been to space until Yuri Gagarin and then Alan Shepard traveled there in April and May 1961. All the technology needed to go to the moon was invented between 1957 and 1969, when US astronauts Neil Armstrong and Buzz Aldrin walked on the moon for the first time. It was an utterly amazing accomplishment and an enormous financial investment to assemble the minds, talent, science, and engineering to pull it all off.

Can this same kind of moon-mission approach help address global warming and climate change? Can humanity agree on any of the approaches that scientists recommend, or deploy any of the proposed geoengineering techniques, before it is too late? Can humanity handle the unintended consequences that are sure to arise, make adjustments for them, and continue forward? In the current economic and political environment, it seems nearly impossible to imagine a coordinated effort at this scale. Yet somehow human beings, as a species, will have to break through and make it happen. If humanity could become as engaged as the engineers and scientists at NASA were during the Space Race, and if we could work at a moon-mission level of effort and creativity, we might be able to solve global warming. We are clearly capable of it if we can all get on the same page and invest wisely in our planet's future.

RIGHT One proposed geoengineering technique to halt the rise in global temperatures is to launch mirrors into space. These large mirrors would reflect sunlight away from Earth.

THE DOOMSDAY BOOK

PANDEMICS AND BIOLOGICAL ATTACKS

THREAT LEVEL: WORLD

It seems as though no one had ever really considered how the HVAC system in a large building or venue works. For example, a big office building, indoor stadium, or convention center has two requirements. First, the air must be comfortable in every part of the structure. This is achieved by circulating heated or cooled air through ducts. Second, fresh air from outside has to be brought in to avoid the buildup of carbon dioxide and the depletion of oxygen from human respiration. The 70,000 shouting fans in a big indoor stadium release a surprising amount of carbon dioxide.

This means that there are two points of vulnerability for introducing pathogens into a building: the place where the indoor air is filtered during recirculation, and the place where outdoor air is filtered before it comes into the building. These filters are serviced and replaced regularly by low-level maintenance personnel, and the plenums that contain the filters are, in many cases, not especially secure.

These simple facts explain how terrorists were able to infect nearly everyone who attended the Super Bowl (about 100,000 people), the Consumer Electronics Show (about 175,000 people), Sundance (about 50,000 people), and a giant UN conference in January, along with similar events in other countries. The terrorists were easily able to gain access to the filter banks in these different venues through bribes or impersonation or by simple locksmithing. Then they sprayed their engineered virus into the venue's airstream. Once the virus became airborne in the buildings, people started inhaling the virus particles and getting infected.

The members of this international bioterror plot had started with a known strain of avian flu called H5N1, which researchers at several labs had started tinkering with around 2010. H5N1 would not normally spread human-to-human, so the terrorists had followed the published work of legitimate scientists and turned H5N1 into a human virus. Further tweaks by the terrorists made it much more contagious than a normal flu strain, even before any symptoms appeared. Then they delayed the onset of

RIGHT While researching a virus like SARS-CoV-2, images from scanning electron microscopes can capture virus particles as they emerge from a cell. Here, SARS-CoV-2 particles are highlighted in gold.

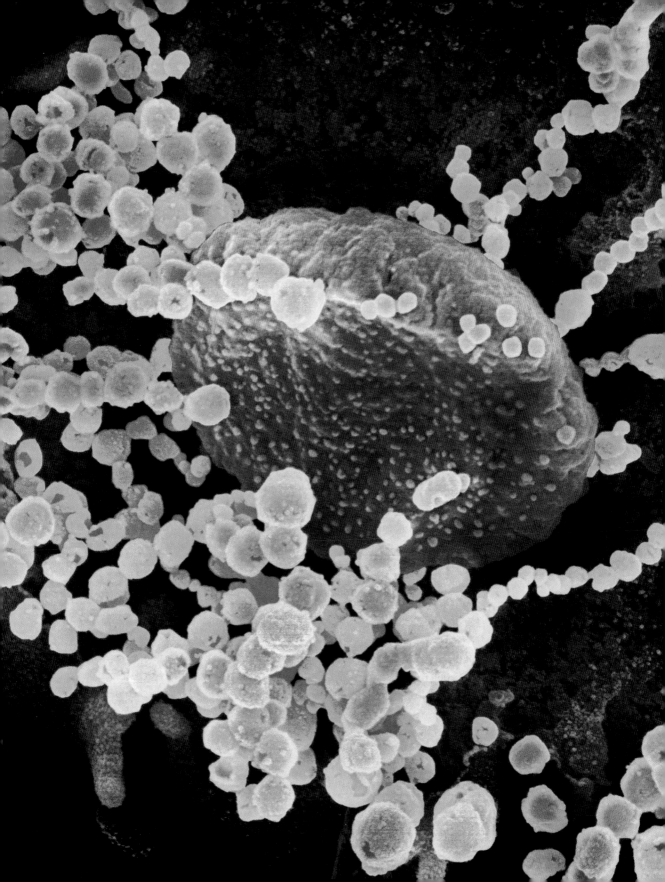

symptoms and increased the survivability of their virus in the open air. Finally, they borrowed a little juice from COVID-19's playbook to increase lethality.

All of the initial infections came when people attended the compromised events in January. Of course, no one knew any of this was happening while the attack was under way. It was only in February that the world realized the grim truth: that hundreds of thousands of people had been infected at these various events, and then these people had spread out around the world while they were still asymptomatic to infect others well before anyone started to die from the disease.

This new, engineered strain of H5N1 had proved to be highly robust in the air and highly contagious, and it killed 10 percent of those who were infected.

By the time the authorities realized what was happening, the only thing they could do was to quarantine sick people and keep healthy people locked in their homes so they could reduce their probability of being infected. Eventually, hundreds of millions of people died in the outbreak before the virus was contained and the pandemic burned out.

SCENARIO

There are two routes to a pandemic. The world got up close and personal with the first route during the COVID-19 pandemic that started in 2020. Here, a naturally occurring virus in animals evolved so that it became infectious to humans and crossed over. Human viruses also evolve regularly, as we see with the flu every year.

The other route is for someone to create and/or release a biological agent. Back in 2001, anthrax spores were mailed to senators and news outlets in the United States. Bioterror scenarios like this frequently appear in movies, novels, and TV shows. In Tom Clancy's book *Rainbow Six*, Clancy imagines a group of terrorists who plan to use a modified strain of Ebola to kill billions of people.

The advantage of a biological agent as a terrorist weapon, as opposed to a chemical agent (see Chemical Attacks and Accidents, page 80), is that biological agents are contagious. One infected person can infect half a dozen other people, and each infected individual can spread the disease on to another half dozen people, and so on. In the case of Clancy's book, the virus was meant to be initially released at the summer Olympics so that its international attendees could quickly spread the virus around the globe.

This is an important factor in the spread of disease in today's world. Every day, there are thousands of international flights transporting people all over the planet. It is quite easy for a pathogen to spread from one country to another undetected. But spreading pathogens internationally is not a new phenomenon—humans have used biological weapons and transmitted diseases in the real world for centuries, both accidentally and deliberately. These weapons have proved to be highly effective.

It is with smallpox, and another enteric fever called cocoliztli, that we see the most important historical cases of accidental, and then deliberate, biological warfare. When European settlers arrived in North and South America starting in the 1500s, it is thought that there were approximately 60 million people living on the two continents. These 60 million people had never been exposed to the variety of illnesses that Europeans had been dealing with for millennia. When Europeans landed in the New World, they brought with them these diseases, which then spread rapidly. The graph on on this page, showing the decimation of the indigenous population in Mexico by disease, is heartbreaking. This is one of the most catastrophic human pandemics that the world has ever seen.

If bioterrorists can engineer, or nature can evolve, a pathogen that replicates some of history's most deadly pandemics, the results could be equally catastrophic today. For example, an especially virulent and lethal virus could have a staggering body count. Even a much less lethal threat like COVID-19, where the body count is tiny compared to something like the Spanish flu in 1918, can cripple the world economy for a period of time, with devasting economic effects on billions of people.

How might a pathogen like this spread? The Allies and the Japanese during World War II show one approach for biological warfare. The Allies stockpiled botulism toxin, anthrax, and brucellosis, though they were never deployed. The Japanese did the same kind of thing and also bred and released fleas carrying bubonic plague. Japan is thought to have killed thousands of people in China with this latter technique. The Japanese even created a plan to infect the West Coast of the United States with bubonic plague using this vector, but it was not carried out because Japan surrendered before the date of the attack. Another approach for spreading disease is to simply gather humans together densely and let nature do its thing. Cholera spreads quickly when human fecal matter contaminates a water supply in a slum. Norovirus starts as a food-borne illness and then rapidly spreads between people—for example, on a cruise ship.

Humans don't have a monopoly on pandemics. It is also possible for pathogens to wreak havoc by

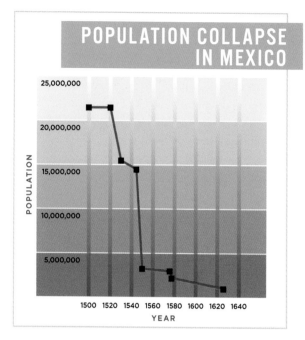

POPULATION COLLAPSE IN MEXICO

ABOVE In the 1500s and 1600s, outbreaks of highly infectious diseases, such as smallpox and cocoliztli, decimated the indigenous population in present-day Mexico.

infecting crops or livestock. In 2019 Eastern Asia, including North Korea, experienced an epidemic of African swine fever in its swine herd. For North Korea, a significant loss of hogs could cause widespread protein deficiencies and starvation in its population of 10 million people. Farmers can also be economically ruined when their entire flocks or herds have to be killed to prevent the spread of a disease.

Today, we have to worry about four relatively new phenomena that can spread pathogens to humans:

1 As climate change melts permafrost, it can release frozen pathogens such as anthrax.

2 Human activities like the deforestation of rainforests put humans in contact with new, uncharted diseases that can cross over from animals. Factory farms and wet markets, where live animals are housed and butchered for human consumption, can be breeding grounds for pathogens as well.

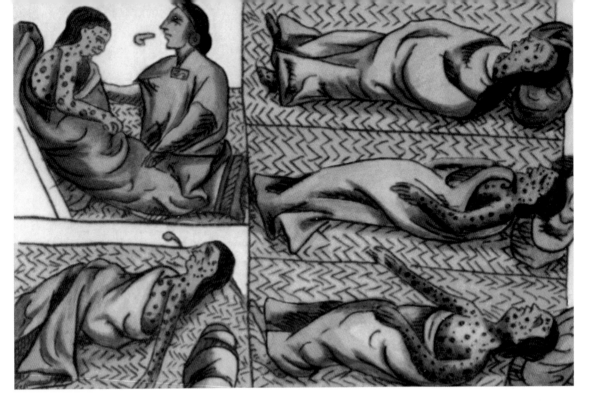

ABOVE Contact between Native Americans and European settlers in North America led to the rapid spread of smallpox among indigenous communities.

3 Existing diseases can mutate to become antibiotic-resistant superbugs or to outwit vaccinations, making conventional cures and preventative measures ineffective.

4 It is becoming easier and easier for individual people to participate in the bioengineering revolution, meaning that one person or a small group could conceivably create and release a bioterror weapon with global implications.

On the doomsday front, the new genetic-engineering technology called CRISPR (clustered regularly interspaced short palindromic repeats) has unleashed the biggest fears. CRISPR makes it much easier to modify the DNA of existing organisms. The field of artificial life has also been advancing, potentially allowing scientists to create their own DNA from scratch to design new life-forms. In the wrong hands, these technologies could create modified pathogens and bioweapons. In 2011, scientists created a mutated form of the H5N1 virus (more commonly known as bird flu) that could spread from person to person far more easily, creating a potentially lethal doomsday agent. At the time, Dr. Paul Keim stated, "I can't think of another pathogenic organism that is as scary as this one."

Using these new techniques, a terrorist organization could in theory re-create smallpox or create a new super-smallpox. Right now, most people are not vaccinated for smallpox, because there is no smallpox in the wild. But there are samples of the virus in various labs around the world. In addition, the genetic material for smallpox (which is tiny, with only 200,000 base pairs) has been sequenced, so there is a digital template for re-creating the virus. With a little bio-tinkering, smallpox could potentially turn into an extremely effective bioweapon.

PANDEMICS THROUGHOUT HISTORY

Bubonic Plague

Humans have experienced outbreaks of the bubonic plague throughout history, but the pandemic that that occurred in the 1300s, better known as the Black Death, is by far the most famous. An estimated 60 percent of the people in Europe (approximately 50 million) lost their lives. Worldwide, counting areas like China and Africa, the total body count was more than 100 million people. This sort of death toll from a disease can be devasting to a country or a region. In 1330, the entire worldwide human population was only about 400 million people, meaning that a quarter of all the people on Earth were lost to the plague.

Why don't we see recurrences of the plague at this scale today? The bacteria that causes bubonic plague is transmitted from one person to the next primarily by fleas. The fleas carry the bacteria from a host animal, such as a rat, to a victim. A person infected with the plague can also cough or sneeze on another person at close range to deliver the live bacteria to a new victim. In the developed world, we have far better controls for rats and fleas, the main carriers of the plague in the 1300s. If someone does contract the plague today, antibiotics can cure it. We still see outbreaks in less-developed parts of the world, though. For example, an outbreak in Madagascar in 2017 affected more than 2,000 people. Groups like the World Health Organization are able to contain these outbreaks with effective treatments long before they become pandemics.

Spanish Influenza

Every year, the flu virus evolves and creates multiple new strains. This is why people need to be vaccinated for the flu annually; the vaccine from one year is ineffective against each year's newly evolved strains. In 1918, a particularly virulent strain of the flu appeared and spread from person to person like

BELOW These "plague cottages" housed those who contracted the bubonic plague during a 1665 outbreak that swept the English village of Eyam. The residents of Eyam agreed to quarantine the entire village for fourteen months, successfully preventing the spread of the disease to nearby towns.

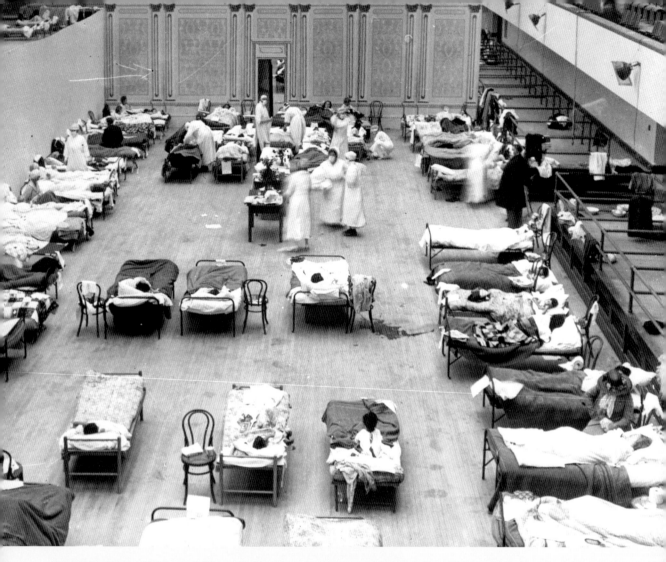

ABOVE During the 1918 Spanish flu pandemic, cities set up temporary hospitals in public spaces, such as the Oakland Municipal Auditorium in California, to help handle the large influx of patients that needed medical care.

wildfire. This strain infected perhaps a third of the world population of humans (about 500 million people were infected), and of those infected, about 1 in 10 died. Fifty million people around the world, or roughly 5 percent of the human population, died from the flu in 1918 and 1919.

Every year, there is an influenza mini-pandemic around the globe that can kill up to half a million people. This pandemic is caused by the constant evolution of the influenza virus, which each year infects hundreds of millions of people, hospitalizes millions, and kills hundreds of thousands across

the planet. Will a flu strain ever again be as devastating as the Spanish flu? There are varying predictions. Today, we vaccinate large parts of the human population for the flu every year. We are also much better at caring for flu patients today, as well as secondary infections like pneumonia that can arise from a flu infection. But if the H5N1 strain were to jump from animals to humans, and if the worst-case scenario occurred where it infected half of humanity with a higher case fatality rate like 5 percent, then scientists estimate that 150 million people could die.

Smallpox

Smallpox is thought to be the deadliest virus humans have encountered. Over the course of human history, hundreds of millions of people have died of the disease. Before its worldwide eradication in 1980, 300 million people had died of smallpox in the twentieth century alone.

Smallpox is fairly contagious (with a R_0 of 4 or 5) and kills about 30 percent of the people who contract it (it also blinds a large number of patients). A person who gets smallpox becomes covered in blisters, and these blisters seep a fluid that contains the virus. Typically, a person with smallpox coughs directly onto another person to transmit the virus, or it is also commonly transmitted when people touch the fluids or scabs from the blisters. This is why clothes and bedding can transmit smallpox virus particles to another person. It is known that a blanket infected with smallpox virus particles can infect people for weeks, and it is thought that collected scabs can remain infectious for years.

HIV (Human Immunodeficiency Virus)

HIV, the virus that causes AIDS (Acquired Immune Deficiency Syndrome), is an example of nature's ability to create new diseases out of the blue.

The virus jumped from chimpanzees to humans in Africa, likely in the 1940s. The virus then spread in Africa for decades before arriving in Haiti and then jumping to the United States. Even though the virus arrived in the United States around 1971 and was actively spreading, it was not detected and did not begin to be characterized until 1981. Scientists did not discover the retrovirus that causes AIDS until 1984.

Why did it take so long? With HIV, it can take years for any problematic symptoms to appear, and the initial symptoms look a lot like those of the flu. These symptoms then often subside for years, even though the patient can be highly infectious to others, before the patient eventually succumbs to opportunistic infections due to a highly compromised immune system. Even then, the symptoms can be all over the map. Many of these opportunistic infections are rare in people with healthy immune systems. It was clusters of these rare diseases that first caught the attention of scientists and allowed them to detect HIV.

Four decades later, there still is no vaccine for HIV. Currently, the best treatments are blends of anti-retroviral drugs that can suppress the virus. Once HIV is sufficiently suppressed, the immune system can function, and the carrier is no longer infectious to others. However, while approximately 40 million people around the world carry HIV today, only 23 million of them are taking anti-retroviral drugs, and millions do not know that they are carrying the virus. Approximately 2 million new people get infected with HIV each year, and about 1 million people per year are dying from AIDS. This situation is a notable improvement over what AIDS patients endured in, say, the 1990s. But a million deaths a year shows that there is still a lot of room for improvement.

COVID-19

COVID-19 (coronavirus disease of 2019) first appeared in the city of Wuhan, China, in late 2019. SARS-CoV-2, the virus that causes it, made the jump from animals to humans. COVID-19 is contagious enough by itself and has one additional feature that makes it hard to stop its spread: carriers can be asymptomatic while being contagious, meaning that many carriers of the disease can walk about without being obviously infectious. Testing is required to detect an asymptomatic carrier.

Although China locked down Wuhan and the entire Hubei province (approximately the size of Nebraska with a population of 58 million), it was too late to stop the new virus from spreading via international travel. Eventually, cases of the virus were reported in nearly every nation on Earth, infecting millions and killing many of those infected, often due to "proinflammatory cytokines in the lungs." In other words, an overreaction by the human immune system inadvertently destroys lung tissue to such an extent that the patient dies.

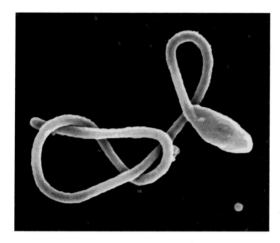

ABOVE This scanning electron micrograph shows an Ebola virus particle.

When we talk about the flu, smallpox, HIV, or Ebola virus, what are we talking about? What is a virus?

A virus is not a cell. A cell is a living entity that can eat and reproduce on its own, whereas a virus particle is not living because it cannot do these things on its own.

To reproduce, a virus particle usually docks onto the side of a living cell and injects its genetic material into the cell. This viral genetic material hijacks the cell's internal apparatus, so the cell starts to create new virus particles. Then, either the host cell bursts and dies, releasing the new virus particles, or the host cell sheds virus particles through the cell wall. When a person has a cold, it means that virus particles are busy replicating inside the patient's cells—in this case, cells in the mucous membranes lining the throat and sinuses. The cells burst and release clouds of new virus particles into the throat and sinuses to infect more cells.

How do viruses spread between people? With colds, these new virus particles enter the air in droplets when a sick person coughs or sneezes. These particles can be inhaled by someone else. The patient also transfers virus particles onto his hands when he wipes his nose. From there, they can move onto doorknobs, refrigerator handles, and countertops. These particles can survive for 12 to 24 hours out in the open. The next person who touches a contaminated object picks up the particles and introduces them, in one way or another, to her nose or mouth. And now another person has a cold. Some viruses, like HIV, are much more fragile and cannot survive in the open. Others, such as smallpox, can exist in the air to a limited degree but can remain infectious on certain materials, like the cloth in a blanket, for weeks.

Living bacteria can also be infectious biological agents. The bubonic plague is an example. While antibiotics can treat bacterial infections by killing infectious bacteria cells, they have no effect on virus particles. In many cases, the only thing that can stop a replicating virus is a patient's own immune system. However, it takes time for the immune system to detect the virus and figure out how to eliminate it. A vaccine is a way to expose the immune system to a virus—usually a dead or weakened form—before the real virus actually arrives. The immune system then learns to create antibodies before they are truly needed. This way, the immune system can immediately swing into action and eliminate the real virus as soon as it is detected.

Some viruses mutate easily—for example, influenza—and people must get vaccinated every year against the newly arising strains. Sometimes new strains outmaneuver the best-laid plans of

ABOVE Creating and manufacturing a vaccine is one major challenge when fighting a pandemic. There also needs to be a supply chain for the supplies needed to administer it, such as syringes.

vaccine designers. In the case of the flu, scientists are hoping to exploit a new breed of vaccine based on DNA and RNA fragments and to find segments of genetic material that are common in every influenza strain and potent enough to create an effective vaccine. If they can discover this Holy Grail, influenza may finally recede as a problem for humanity.

Antiviral medications have also emerged as a relatively new form of treatment. These medicines can slow down the reproduction of new viral particles, reducing symptoms. Antiviral drugs are especially popular in treating the flu. A typical antiviral prevents a virus particle from docking with a new cell or injecting its genetic material once docked.

THE PARAMETERS OF AN INFECTIOUS DISEASE

A disease's lethality in a population is determined by two things: the rate at which it can spread from person to person, known as R_0, and its case fatality rate. The Ebola virus can be especially lethal because it is highly contagious (R_0 of 4 or 5) and also because it kills about 90 percent of the people who are infected with it. Meanwhile, measles is much more contagious (R_0 of 15) but with a much lower case fatality rate (less than 1%). Other factors at play are the incubation period (the time between infection and the appearance of symptoms) and the length of time for which an infected person sheds the pathogen and is able to infect others. One unfortunate trait of some diseases is asymptomatic shedding, where a person who looks healthy is shedding virus particles or bacteria, unknowingly infecting others. Asymptomatic shedding is especially common with COVID-19.

PREVENTION

If a terrorist organization were to launch a biological attack, or if a disease from nature begins rapidly spreading, what should the affected countries do to respond?

One approach is to quickly detect a biological attack or outbreak in its early stages. In the 2001 anthrax attacks, five people died from these attacks, and there was widespread concern that larger-scale anthrax attacks would happen. To counter these concerns, the United States deployed the BioWatch system in a number of major cities. The idea is to suck ambient air into filters and analyze them to look for infectious agents (such as virus particles, bacteria, and spores). Current technology allows detection of a biological agent within 24 to 36 hours.

In concert with BioWatch, a coordinated system to report suspicious cases in both humans and animals allows anything unusual to be noticed and prioritized quickly. For example, if a patient were to visit an emergency room with the symptoms of an anthrax infection, doctors could use this reporting system to quickly notify the country's whole health apparatus that an attack or an outbreak might be under way. The Centers for Disease Control handles information gathering and dispersal to the entire health-care community in the United States.

For containing the spread and treating a new infectious agent once it's discovered, a country needs to be prepared. During the COVID-19 pandemic, few Western countries had good responses at the outset. As a result, the United States, Spain, Italy, France, Germany, the United Kingdom, and others saw thousands of deaths. But other places like Taiwan had their playbooks dialed in, kept their per-capita death rates very low, and provide lessons in preparedness:

1 Make widespread testing available to everyone in the country to catch every new infection, including those who are asymptomatic.

2 Immediately quarantine everyone who is infected.

3 Use contact tracing to track infected people and the people with whom they come in contact. Some countries have started using apps to help with this process and make it automatic.

4 Require an entire population of people to practice social distancing for a period of time to reduce the spread of infection. By asking everyone to stay home and avoid gatherings, the pathogen can lose momentum, especially if there is also widespread testing and quarantining. During the COVID-19 pandemic, several countries found significant benefits from mandating people to wear masks in public.

5 Create stockpiles of essential medical supplies and equipment ready for instant access and ramp up domestic suppliers and supply chains to satisfy the intense demand created by a pandemic.

6 Implement travel restrictions and mandatory quarantines to prevent infected people from entering and spreading the disease in a country.

7 Create education programs to help the public understand how to avoid infection.

8 Increase research and manufacturing capacity so we can rapidly develop and deploy vaccines (for viruses) and antibiotic cures (for bacteria).

9 Understand the economic implications of a pandemic for a nation's—and the world's— economy.

During the COVID-19 pandemic, many countries learned valuable lessons about the importance of medical preparedness. As the disease spread, there was an immediate worldwide need for personal protective equipment (PPE) for medical personnel (including N95 respirator masks), ventilators, and medical supplies and sedatives for intubated patients. Because demand for ICU beds was projected to overwhelm existing supply, field hospitals needed to be built quickly and stocked with personnel and supplies as well. Countries like the United States and Japan learned about their strong dependence on China for critical medical supplies like antibiotics and masks and then were unable to obtain these essential items as China's production capacity shut down during its COVID-19 outbreak. The concept of a globalized economy that relies on a "just in time" production strategy came under scrutiny.

Besides the slow response on procuring medical supplies, several countries also suffered from a severe lack of test kits. Unlike Taiwan and South Korea, which were able to contain the virus quickly because they could initiate widespread testing, the United States suffered from a lack of test kits and slow testing for months after the virus arrived. This meant that the spread of COVID-19, especially through asymptomatic carriers, could not be monitored or abated. Had the United States been able to test everyone and definitively quarantine all those infected (including asymptomatic carriers), the impact of the virus, the duration of the outbreak, and its economic effects would all have been much less severe.

To understand why quarantine and education for the public are important, let's take Cuba's response to HIV in the 1980s and 1990s as a case study. Cuba's response was far more aggressive than

BELOW A burial team puts on personal protective equipment before assisting with the burial of an Ebola victim. Because the Ebola virus can be easily transmitted through infected body fluids for days after an infected person has died, safe burials have likely saved thousands of lives.

what we saw in most other countries. Any Cuban found to have the AIDS virus was placed in mandatory quarantine in a sanatorium where carriers were given a wide-ranging education in the disease they were carrying and how they could prevent its transmission. This mandatory quarantine was later relaxed, but new patients still need to enter an eight-week educational quarantine before they are released. If HIV carriers are found to have practiced unsafe sex after finishing the eight-week program, they are quarantined permanently.

Did this approach work? Absolutely. Though there are questions about the ethics of mandatory quarantine and the feasibility of implementing these measures in other countries, many lives were saved in Cuba. In the United States, thirty-five times more people per capita died of AIDS than in Cuba.

What else can we do, especially in the case of intentional bioterrorism? Author and entrepreneur Rob Reid has spoken widely on biological terror and synthetic biology. He advocates a multi-pronged approach. His prescription for creating a global

THE SIDE EFFECTS OF SOCIAL DISTANCING

The response to the spread of the COVID-19 virus made social distancing the new norm around the world. Depending on the country, the citizens were either told to or forced to stay home with the primary goal of "flattening the curve." Every nonessential enterprise where people could congregate was closed, including restaurants, gyms, arenas, stadiums, schools, universities, parks, conventions, trade shows, churches, and movie theaters. In comparison to the flu, COVID-19 had a larger R_0, was more lethal, and could spread asymptomatically. Letting people go about their daily lives would have caused an intense spike in the number of cases. Social distancing helped prevent this spike, which would have overwhelmed health-care systems potentially and killed millions unnecessarily.

The unfortunate side effect of social distancing was a brutal slowdown of the economy. Hundreds of millions of people around the world became instantly unemployed because businesses closed. Supply chains were also affected. Some products became difficult to purchase because household demand for goods spiked while commercial demand crashed during the lockdown. There were toilet paper shortages, as tens of millions of people who had been using the restrooms at school and at work suddenly switched to home use. The same happened to many food items as home consumption of food grew and restaurants closed.

While the virus is spreading rapidly (before herd immunity starts to kick in or a vaccine gets developed), there is a difficult tradeoff between the economy, the medical capacity to treat cases, and the death toll. Forcing everyone in a nation to stay home has a gigantically negative economic impact. But with COVID-19, it was the only tool available to prevent a spike in cases from overloading the hospitals and causing multitudes of unnecessary deaths. In a similar future event, it would be ideal if everyone in a country could be regularly tested, every carrier could be immediately quarantined, and contact tracing could identify any encounters between infected and uninfected people. In this ideal case, no stay-at-home orders would need to be issued.

"immune system" emphasizes the importance of embracing synthetic biology technology rather than demonizing it. The potential for good is equal to or greater than the potential for evil. By letting the existing experts in synthetic biology teach us and accelerate the development of the technology, and by creating more experts, the technology can advance more quickly to help more people. In addition, these experts can help fight synthetic pathogens that terrorists create.

Reid also recommends actively building defenses against bioterror weapons. We have the potential to create advanced biosensors to detect pathogens quickly and effectively and to make these sensors as ubiquitous as smoke detectors., We could develop the ability to rapidly create and test vaccines and antibiotics and even invent printer technology to manufacture vaccines and medicines in pharmacies and homes so that billions of vaccine doses can be instantly available.

Michael Leavitt, former secretary of health and human services, famously noted: "Everything we do before a pandemic will seem alarmist. Everything we do after will seem inadequate." Having witnessed the global effects of COVID-19, it is time for the world to seem alarmist. Humanity needs to preemptively think about and build effective solutions to the threat of highly contagious and highly lethal diseases. Pandemics, whether from nature or engineered by humans, are only going to become more prevalent in the future. Because we know for certain that more pandemics are coming, we should take aggressive steps now to prepare for and eliminate this threat.

RIGHT During the COVID-19 pandemic, normally bustling city streets and tourist destinations, such as Times Square, were ghost towns, as mandates to stay at home and practice social distancing around the world took effect.

CHEMICAL ATTACKS AND ACCIDENTS

THREAT LEVEL: CITY

It is impossible to believe, but the chemical attack is under way as we speak. The air-raid sirens are blaring all around the city. I am sealed in my safe room and I am breathing through my gas mask, and I am hoping to survive this attack. The whole thing is surreal, like I am a character in a disaster film. But this is as real as it gets.

Here is what we know, or at least what we are told. We are told that there are hundreds of Scud missiles on their way. These ballistic missiles were developed in the Soviet Union and sold to Iraq. These missiles have a simple, reliable design that allows them to fly hundreds of miles. We are told that multiple Israeli cities are being targeted. We are told that the missiles have reasonably accurate guidance systems; once launched, the missiles can land within a mile or two of the intended target. While these missiles can't take out a specific building, we can be reasonably sure that the missiles will land inside the city, somewhere.

And here is the most terrifying part, and the reason why I am in my safe room sealed tight with plastic and tape, wearing a gas mask: these Scud missiles can carry chemical-weapon warheads. A single missile can carry a metric ton of VX, a synthetic nerve agent, or a metric ton of mustard gas, a powerful blistering agent, into the heart of the city. Just a tiny amount of VX is needed to kill a person.

It also sticks around. If someone touches an object contaminated with VX, they can absorb it through the skin and die. When inhaled, mustard gas blisters the lungs, leading to suffocation.

The country took the threat seriously and prepared everyone for this potential disaster. It manufactured and issued millions of gas masks and antidote injector pens. It distributed plastic and tape to create our safe rooms. And it educated us so we would be prepared when the time came. As the sirens started blaring, we all knew what to do. We picked a room and taped plastic over the windows, doors, and vents as a way of keeping the gas from seeping in.

All we can do now is sit in our safe rooms, wearing our masks, and pray.

RIGHT This photo from 1997 shows steel containers of VX, an incredibly deadly nerve agent, stored in a facility at Newport Chemical Depot in Indiana. The United States plans to finish destroying any remaining VX stocks by 2023.

ABOVE This scene from World War I shows soldiers tending to a man who has been exposed to gas.

The scenario described above is inspired by real-life fears. During the Gulf War in 1991, Iraq routinely attacked Israel with Scud missiles launched from western Iraq. Although Iraq did not use chemical weapons against Israel, it was widely believed at the time that they would do so.

A large-scale chemical attack or an industrial accident is a nightmare doomsday scenario. Whereas a conventional bomb affects only the immediate vicinity around the site of its explosion, the poison in a chemical weapon can spread much farther, potentially injuring and killing many more people. There are several "advantages" of a chemical weapon from a military or terrorist standpoint:

- Chemical weapons are relatively inexpensive.

- They are relatively easy to manufacture or obtain (especially compared to something like a nuclear weapon).

- They are more potent than a conventional weapon, pound for pound. As we will see later,

just a gallon (4 l) of a nerve agent like VX can kill tens of thousands of people.

- Some chemical weapons can contaminate a site for a long period of time.

The first recorded use of chemical weapons goes all the way back to 600 BCE, when soldiers from Athens poisoned the water supply of one of its enemies. But the use of chemical weapons in any significant way did not occur until World War I.

World War I was a game-changing event in terms of military technology. It was the first war where machine guns, tanks, and airplanes were used in a widespread way. It was also the first war to see the widespread use of chemical weapons. The deployment of mustard gas, phosgene gas, and chlorine gas during this conflict taught the world how dangerous chemical weapons could be. There were over a million casualties caused by chemical weapons in WWI. The world was so horrified by the effects of chemical warfare that the Geneva Protocol banned the use of chemical weapons in 1925.

A large-scale chemical attack can potentially kill tens of thousands of people if there is enough of the chemical and enough people in the vicinity. And it does not need to be a weapon specifically. There are many possibilities for industrial accidents, or intentional industrial attacks, that release significant amounts of chemical toxins into the surrounding area. The Bhopal disaster (see sidebar on page 84) is an example of how badly an industrial accident can go wrong. What we see in the Bhopal incident is the typical signature of a large-scale industrial accident. First, there is really no way to control where a gas cloud goes, nor is there a way to stop it. The prevailing winds and topography of an area control the path of the gas. Unless there is an intricate plan in place ahead of time, there is no way to warn civilians in the path of the cloud and nowhere for them to go. Emergency personnel are not going to want to enter the area until the main danger has dissipated, and they would need to suit up in hazmat gear and respirators, limiting the number and duration of personnel who can enter the area. By then, the damage would be done.

CHEMICAL WARFARE

Efforts to develop chemical weapons started in World War I and initially involved the use of these three different chemicals:

- **Chlorine gas:** Chlorine gas is relatively inexpensive and easy to produce. To create chlorine gas, salt (chemical composition: sodium chloride, or NaCl) and water are combined to create a brine. Using electrolysis, the chlorine is easily separated and released as a gas. When chlorine gas gets in the lungs, it mixes with moisture and forms hydrochloric acid (among other things), which causes lung damage and leads to choking. Chlorine gas can be made in abundance easily and at low cost. The disadvantages are that chloride is a gas (and therefore takes up a lot of space, even when compressed), and a large amount of chlorine is required for it to be effective. The first successful chemical attack in World War I used chlorine gas. In 1915, Germany made over 100 tons of the gas, stored it under pressure in nearly 6,000 large gas cylinders, and moved these cylinders to the front lines. Seven thousand soldiers were injured and 1,000 died when the Germans released and the breeze carried the heavier-than-air gas into the trenches of the opposing forces. It has been used as recently as 2007 in a number of small attacks in Iraq. Today, one new fear with chlorine gas is that terrorists could release it by using an attack on rail cars transporting the gas or on storage tanks in a populated area. Over 10 million tons of chlorine gas are created and shipped in the United States every year.

- **Phosgene gas:** Like chlorine gas, phosgene gas is a choking agent that is easy and inexpensive to make, but it is far more lethal. To manufacture it, a factory mixes chlorine gas and carbon monoxide together in the presence of a catalyst to form $COCl_2$. Like chlorine gas, phosgene gas is stored in many pressurized cylinders, making it fairly cumbersome to use on a battlefield. The effects of phosgene gas are also delayed, meaning a soldier exposed to it does not immediately feel any effects and can keep on fighting. Even so, the large majority of chemical warfare deaths in WWI are attributed to phosgene gas.

- **Mustard gas:** Mustard gas is a liquid mist that contains a form of sulfur that acts as a blistering agent. Mustard gas affects the lungs, eyes, skin, and any mucous membranes, causing huge fluid-filled lumps that often get infected. In the lungs, mustard gas impedes the body's ability to process oxygen by blistering delicate lung tissue. Longer-term, mustard gas is a carcinogen because it damages DNA.

MODERN-DAY INCIDENTS

There have been a number of examples of chemical attacks, both accidental and intentional, over the last century.

The Bhopal Disaster

In 1984, a Union Carbide plant in Bhopal, India, leaked 40 tons of methyl isocyanate (MIC) gas into surrounding neighborhoods. Over half a million people near the pesticide plant were injured, and nearly 20,000 people ultimately died from their injuries. The gas was created when water accidentally entered a tank at the plant containing 42 tons of liquid MIC. The water and MIC combined to create a very powerful, hot reaction that turned the liquid MIC into a ground-hugging gas, blown by light winds into the town of Bhopal at approximately 1:00 a.m. MIC is highly toxic because it causes extensive lung

BELOW Ruins of the Union Carbide plant, where an accident in 1984 exposed residents in Bhopal, India, to MIC gas and killed thousands.

damage when inhaled. The lungs then fill with fluid, causing the victim to suffocate. MIC can also be absorbed through the skin. This incident is still considered to be the world's largest industrial accident.

The Lake Nyos Disaster

Even nature can cause chemical attacks. A very bizarre form of natural chemical attack occurred in Cameroon, near Lake Nyos, in 1986. This lake is unusually deep, more than 300 feet (91 m) in some places. There are high concentrations of carbon dioxide gas that are trapped in the deeper sections of the lake. In the case of Lake Nyos, the carbon dioxide comes from volcanic magma that exists beneath the lake. The carbon dioxide dissolves in the deep water, and the high pressure at those depths holds the dissolved carbon dioxide in the water in the same way that a sealed can holds dissolved carbon dioxide in soda. An unknown event caused the lake to invert, pushing much of the deep water to the surface. With the pressure released,

the water effervesced and thousands of tons of carbon dioxide were released into the air. Since carbon dioxide is heavier than air, this massive amount of carbon dioxide gas clung to the ground, displacing the oxygen and suffocating thousands of people and animals in the vicinity up to 14 miles (22.5 m) from the lake. Today, there is an ongoing effort to de-gas the lake using pipes that reach down to the deepest waters in the hope of preventing a repeat.

Halabja Attack

In 1988, the government of Iraq under Saddam Hussein bombed the Iraqi town of Halabja with a combination of mustard gas and nerve toxins. Approximately 5,000 people died, and perhaps twice as many more were injured as well. The attack is historically notorious and considered an act of genocide, because the Kurds who lived in the town were primarily civilians. The attack left a legacy of cancer and birth defects for those who survived.

Tokyo Sarin Gas Attack

Sarin became famous because of several sarin gas attacks by members of a Japanese doomsday cult known as Aum Shinrikyo. The cult established a

ABOVE After Lake Nyos released a destructive amount of carbon dioxide in 1986, the lake turned brown. When the lake inverted, it also brought sediment with a high iron content to the surface.

manufacturing facility for sarin. They then used the sarin in a series of gas attacks to kill dozens and injure thousands of people. The most widely known attack occurred when the cult released sarin gas into a subway train in Tokyo in 1995. As many as 6,000 people in the subway system may have been affected.

Moscow Theater Hostage Crisis

In Moscow, Chechen rebels stormed a Moscow theater in 2002 and took 850 people in the theater as hostages. Authorities pumped an aerosol containing carfentanil into the ventilation system to gas the attackers. Unfortunately, the hostages were gassed as well. As a result, all forty attackers and 200 of the hostages died. Carfentanil, invented in 1974, is an opioid that is approximately 100 times more powerful than fentanyl (which is 100 times more powerful than heroin). Because of its potency, some countries now classify carfentanil as a chemical weapon.

To create a doomsday scenario with a chemical weapon, the chemical agent used in the weapon needs to be concentrated and lethal. Military scientists have spent years searching for the most potent chemicals possible.

The thing that changed the landscape of chemical warfare was research on effective insecticides. When DDT was created, it revolutionized the control of insects, but it also had significant environmental problems. Further study into more environmentally friendly insecticides with fewer long-term side effects made it possible to imagine very effective chemical poisons for people. These chemicals are known as "nerve toxins" or "nerve agents." It takes only a very small quantity of a nerve agent to kill a human being.

Any neurotoxin works by disrupting the normal flow of signals in the nervous system. For example, you may have heard that eating certain parts of a puffer fish can be fatal. This occurs because parts of the puffer fish, such as its liver, contain a neurotoxin called TTX (tetrodotoxin). This chemical finds its way into nerve cells and blocks the cells' ability to process sodium. Eventually the cells stop working, and muscles like the diaphragm no longer receive the nerve signals that control breathing. The victim suffocates.

Other nerve toxins produce similar effects by overstimulating nerve cells. When you spray a bug with a can of household insecticide, you are often hitting the bug with a nerve agent. If so, you may see the bug start twitching. Bugs and humans both use a neurotransmitter chemical called acetylcholine

BELOW The search for more effective pesticides spurred the development of chemical weapons as many of the chemicals used to kill pests can also cause death in humans.

to allow a nerve cell to trigger a muscle contraction. Once the muscle fires, a second chemical called acetylcholinesterase breaks down the acetylcholine. Many nerve agents act to block the action of acetylcholinesterase, so the muscle keeps contracting. Eventually the muscle gets so fatigued that it stops functioning. If it happens to be the diaphragm muscle, the victim suffocates.

One example of a nerve agent that was discovered while researching pesticides is sarin. Sarin affects the nervous system, causing loss of muscular control. Sarin is an effective poison because it can be easily vaporized. Breathing in a tiny amount of sarin vapor, or exposing skin to sarin liquid, causes death very quickly. Because sarin comes in the form of a liquid, it is easier to transport. Sarin is twenty-six times more lethal than cyanide gas, and 100 kilograms (220 pounds) of sarin (which would easily fit in a car) is enough to kill a million people.

VX is another example of a nerve agent that was first developed as a potential insecticide. Ten milligrams of VX is enough to kill a human being in a few minutes. VX is even more potent than sarin—a mere 10 kilograms (22 pounds) could theoretically kill a million people. Like sarin, VX causes its victims to die by suffocation by disabling the muscles that control breathing. VX, like carbon dioxide, is heavier than air, so it tends to stay at ground level.

One other development in the VX saga is VX-2, known as a binary agent. Because VX is so deadly, it can cause death to the attackers. A binary agent consists of two separate chemicals that are relatively benign but, when mixed, react to form the deadly compound. VX-2 is therefore much safer to handle, when, for example, it is being loaded into a missile. The two compounds mix while the missile is in flight to create VX.

As seen with the Moscow Theater Hostage Crisis (see page 85), synthetic opioids such as fentanyl and carfentanil can also be turned into chemical weapons that are as powerful and deadly. Doctors prescribe opioids medically because opioid molecules dock with nerve cells in the brain and nervous system; once docked, they turn down the volume on pain signals. This pain-relieving action can help a patient suffering from chronic pain (caused by cancer, for example) or intense pain after an injury or surgery (see page 116).

Unfortunately, opioids also suppress the respiratory system. This side effect is the primary cause of most opioid-related deaths. When a person overdoses on opioids, it can lead to complete respiratory arrest—the victim stops breathing—and unless there is an immediate intervention, the victim will suffocate and die very quickly.

Only a tiny amount of fentanyl, measured in micrograms, is needed to provide the analgesic effect for which it is medically intended. Unfortunately, just a slightly larger amount, and here we are talking about 2 milligrams, is fatal for a typical adult. This tiny amount will shut down the respiratory system, suffocating the victim in minutes. For comparison, a single grain of table salt weighs approximately a milligram.

Fentanyl can be absorbed through the skin or inhaled as a dust. If some object, such as a countertop or doorknob, is dusted with fentanyl, it is possible to absorb a fatal dose through the skin. Carfentanil is a hundred times more powerful than fentanyl—an even smaller amount can cause death. In theory, a pound (0.5 kg) of carfentanil could kill a million or more people if spread as a dust or a mist in a large crowd.

PREVENTION

If we sit down and contemplate all the ways that an enemy can kill tens of thousands of people with a chemical attack, we can begin to understand the scope of the problem. Here are just a few examples:

- An attacker can carry a canister of chemical into a venue (train station, auditorium, convention site, theater, arena, etc.) in a backpack or suitcase and release the chemical.

- An attacker can release a chemical into a building's ventilation system.

- A large quantity of a chemical agent can be transported into an area with a car or truck and then either detonated or sprayed.

- An attack on a storage facility or transport vehicle for a lethal industrial chemical could release chemical agents in a dense urban area.

- At a large music festival or outdoor stadium, an airplane or a swarm of drones could be used to spray a chemical agent onto the crowd.

One effective approach is deterrence. Ventilation systems are such an obvious point of attack that they should be made impenetrable in terms of security. The airspace over large-scale events is another vulnerability. Preventing airplanes from overflying an event is relatively straightforward if considered ahead of time. This can be expensive because it involves using fighter jets to patrol the area, but it is worth the expense when large groups are gathering. Keeping drones out is another matter. A drone defense system to protect large events is one obvious technique to consider (see Drone Strikes and Swarms, page 12). Securing chemical storage facilities at industrial sites is another obvious step, while securing rail lines that transport industrial chemicals is much more difficult.

We can also preemptively thwart plans for a chemical attack. This approach hinges on intelligence gathering and spying and then acting on information being collected. One opportunity for a government to deter terrorist attacks is to focus its efforts on ferreting out an organization's or country's intentions and attack preparations long before an attack occurs.

No one wants to live in a society where a government spies on every citizen, but with groups like Aum Shinrikyo acting as proof that deadly chemical attacks are possible, we may need to take steps in this direction, using an approach that is as unoppressive as possible. Being able to detect intent early would allow us to thwart many terrorist threats, including chemical attacks, and would also help us prepare for other doomsday scenarios such as pandemics (see page 66). In 2019, the UN announced plans to deter terrorism by, among other things, rapidly deleting terrorist content from the internet, tracing terrorist activities and communications, and uncovering sources of financing for terrorists. Perhaps an impartial, omniscient AI that is listening to everything is a direction to explore.

Another way forward is to aggressively prepare for attacks. We can learn lessons about preparation from Israel. In the 1990s, the Israeli government gave its citizens the following items to use to protect themselves against chemical attacks:

- **Plastic sheets and tape:** By taping plastic over doors, windows, and vents in a room inside a home or apartment, Israelis could improvise "safe rooms" and prevent a chemical agent from entering the room. This technique is not perfect, but it could go a long way toward minimizing chemical infiltration. Israel later enacted building codes so that apartment buildings and businesses are constructed with more formal safe spaces (for example, with proper chemical-filtering ventilation systems).

- **Gas masks:** Then gas masks were issued to be worn inside safe rooms in case any chemical vapor leaked through the plastic and tape. Israel issued five million gas masks to its citizens and continued to manufacture and issue gas masks until 2014.

- **Auto-inject antidote syringes:** Exposure to nerve agents like sarin and VX can be treated with the drugs atropine and pralidoxime chloride. These drugs need to be immediately available, because there is only a small window of time when they are useful. Once the effects of sarin or VX have progressed far enough, the antidotes are no longer effective. Israeli citizens were issued these drugs with their gas masks.

Implementing these measures in a country with a larger population than that of Israel, where fewer than 10 million people live, would be more challenging. Plus, needing to carry around your gas mask and antidote pens every time you go to an event gets old fast. It will seem like a waste of time—until a large-scale attack happens.

But what alternative is there if chemical attacks pose such a serious threat? What do we do if a dense crowd of a hundred thousand people is gathered for an outdoor festival and a plane flies over, spraying carfentanil? What about the people who live near a plant that handles toxic materials? Or what if thousands of commuters on a train during rush hour are suddenly gassed with a lethal chemical agent? If everyone in the crowd has a gas mask and the proper antidote pens, there is some hope for survival.

As we saw in Pandemics and Biological Attacks (see page 66), Michael Leavitt described how "everything we do before a pandemic will seem alarmist. Everything we do after will seem inadequate." The same principle applies to chemical attacks. Without concrete efforts to prevent them and protect the population, a chemical attack is one of those doomsday scenarios that seem inevitable.

BELOW These masks were produced at the Shanlon Gas Mask Factory, the largest manufacturer of gas masks in Israel.

GRID ATTACK

THREAT LEVEL: COUNTRY

It's always jarring when the power goes out, but it's usually no big deal. This time, we were watching TV when it suddenly turned off, along with the lights, the refrigerator, and the furnace. The background hum in the house stopped, and it seemed a lot quieter than normal. A power failure that hasn't obviously been caused by the weather usually resolves itself in a moment or two. But after a couple of minutes, nothing had changed. We were still sitting in the dark.

This was true darkness—the streetlights outside and the lights in all of the other houses were dead. None of the night lights in the house were working. Since it was already late, we decided to go to bed, assuming that everything would be fine in the morning.

We woke up the next day. The power was still out. Our phones exhausted their batteries because they hadn't recharged overnight, and the internet was dead because the house had no power.

What was going on? We went to the car and turned on the radio. We tuned it to the news on a local station and started to get the scoop. The station was running on generators—it was the only reason they were still on the air. There had been a large-scale cyberattack on the power grid that had shut down the power in some regions. That was followed by a wide-ranging terrorist attack on the physical infrastructure of the grid. The combination had wreaked havoc on the entire eastern half of the country. No one had power for hundreds of miles in any direction.

Without power, all schools were closed. Most businesses were closed. Hospitals and a handful of pharmacies running on generators were open. Only a few gas stations were open, with impossibly long lines. City officials were asking us to conserve water. We were advised to treat the day like a snow day and just stay home. With all traffic lights out, keeping cars off the road would prevent accidents.

One day then turned into several. With the power out for a few days, everyone's refrigerators and freezers were warming and the food was melting. People who had a gas stove or a gas grill started cooking food and serving it. The next two days felt like a huge block party, and we were all eating like kings. Neighbors with generators were generous, leaving out power strips so people could charge their phones.

But as three days became four days and four days became five, the mood started to change. Reports on the radio were starting to sound more and more ominous.

The repairs were taking longer than anyone had

RIGHT A fire breaks out at a substation. Fires such as these can lead to widespread power outages.

predicted. The terrorists had chosen their targets carefully, and hundreds of little three-man cells had wreaked havoc over a very wide area.

So much of the critical infrastructure had been damaged by the terrorists that a week without the power grid became two weeks and then three. Without a power grid, *everything* was affected. On a normal day, a grocery store receives trucks full of inventory all day long. Now, because of the lack of refrigeration in stores and the loss of the normal supply chains and warehouse operations, food was getting scarce. Gasoline was in short supply as well, because very few gas stations were open. And it was getting harder to move gasoline around in tanker trucks. Medicine from pharmacies became a big problem too, as supplies ran out. Some cities were having water-supply problems as their equipment failed from lack of power, lack of fuel for generators, or lack of spare parts.

No one had anticipated that a power failure affecting millions and millions of people would last for weeks. The East Coast devolved until it looked almost like a war zone, with supply lines delivering just the basics so that people didn't starve. People descended like locusts on food trucks pulling up with dry goods and cans of food. Many of the most vulnerable people began dying. They couldn't get their medicines, or they were too remote to get food. People started attacking one another with guns as the situation became more desperate. Hungry people will often take extreme actions to get food; hungry people with children even more so.

Just the simple act of paying for things became a challenge. Without power, businesses couldn't process credit cards. ATMs didn't work, and there wasn't enough cash to go around anyway.

No one was able to work. Without the ability to earn money, financial difficulties went through the roof for the vast majority of people. How could we pay our rent or our mortgage? How could anyone go back to school? When all was said and done, this long-term outage set the whole country back by a decade, and more than 3 million people died.

SCENARIO

A widespread collapse of the power grid is a doomsday scenario in any country that is unprepared for it. So many things in daily life—many of them quite important—depend on a constant supply of clean power from the grid. It is easy to forget everything in modern society that needs power, and lots of it. The list of important things that require power includes stores, malls, and shopping centers as well as the factories and warehouses to create and deliver their products; all businesses, including banks, gas stations, and restaurants; doctors' offices, hospitals, and medical devices, especially if they are used at home; and security systems, payment systems, ATMs, traffic signals, and street lights.

For the average person in a developed country today, a power outage can take away their ability to work, pay bills, shop, entertain themselves, buy and cook food, refrigerate and freeze food, get routine medical care and medicine, travel, heat and cool their homes, light their homes, and secure their homes for more than a day or two. A person's whole way of life collapses.

Even worse, if the grid collapses across a region with a large population, the entire economy can collapse as well. Since many working people, along with their children, are living from paycheck to paycheck (more than 50 percent of all Americans do), they rapidly run out of money if they cannot work. Likewise, businesses rapidly fail if they cannot open. Once the power comes back on, it does not magically undo the damage the outage has caused.

And a bigger set of problems looms—problems like looting, riots, food panics, and deaths from lack of medical care and medicine. The fabric of society can break down in weeks without power. Sometimes there is a power failure and things remain calm. At other times, we have seen power failures go badly. Three important examples include:

1 **The New York City Blackout of 1977:** Vandals attacked thousands of properties and lit hundreds of fires, 3,000 people were arrested, and hundreds of police officers were injured in the process. The blackout only lasted a day, but damage estimates were $350 million ($1.5 billion in 2020).

2 **Venezuelan Power Outages of 2019:** During a nationwide power failure, looters attacked downtown Maracaibo and hit hundreds of businesses, as roaming crowds caused chaos. Even armed guards were forced to retreat.

3 **Hurricane Maria Power Failure in Puerto Rico:** Hurricane Maria in 2017 caused an island-wide outage. There was some looting and chaos, but the larger issue was the thousands of deaths that resulted, as the power failure took months to resolve.

Under normal conditions, a small police force can manage the small number of crimes that occur.

In a power failure, there are no streetlights. People are hunkered down in their homes. Without foot traffic or automobile traffic in business areas of town, few businesses are open. With emergency services taxed to the limit, the potential for crime rises dramatically. Breaking the law can become the norm rather than the exception.

HOW HELPFUL ARE BACKUP GENERATORS?

There are three kinds of generators we might see in everyday life: small portable gasoline generators, usually 2,000 to 5,000 watts in size, that you can buy at a home improvement store; big permanent generators like the ones found in a hospital or a large office building; and tractor-trailer–size mobile generators that can appear temporarily at construction sites and concerts.

All three rely on a fuel supply. A typical hospital, for example, would keep a big tank of diesel fuel on site to give its generators four days of operation. The assumption is that either the power failure would be resolved within this time frame, or the tank can be resupplied. But if the power failure lasts long enough, and fuel resupply becomes impossible, then backup generators will run out of power and therefore cannot serve as a long-term solution to a grid failure.

At its simplest level, a power grid is a fairly straightforward organism. To understand a basic working system, let's look at a single city that has its own power plant and power grid. The grid consists of these components:

1 The power plant, which produces all power for the residents and businesses in the city.

2 The transmission lines, which take power from the power plant to the points where it is distributed.

3 The distribution lines, which carry the power through the city and send it off to neighborhoods and then individual addresses.

4 The substations and transformers, which convert power from high voltages to lower voltages and vice versa.

A simple, single-city model like this would work well in most cases, and in fact this is how the power grid got started long ago. But what happens when the power plant needs to go offline for maintenance? To handle this problem, the city might need two or three power plants instead of one. But this option is expensive. To solve the maintenance problem, all power plants in different cities can connect with one another through transmission lines, allowing them to

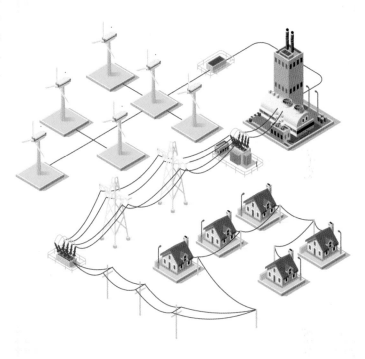

ABOVE A diagram showing how electricity is delivered from a central power plant to individual homes and businesses using a traditional power grid.

LEFT In this photo from March 2019, car headlights provide the only light in Caracaras, the capital of Venezuela, as it experiences a nationwide power failure.

share one another's loads when a power plant has a problem or needs to go offline. This "big interconnected power plant architecture" is the essence of the power grid in the United States and many other countries today.

Why did this approach evolve, as opposed to, say, every home and business generating its own power with a generator on site? For example, every home makes its own hot water instead of getting the water from one plant that generates hot water for the whole city; every home has its own air-conditioning system rather than being cooled by one giant air-conditioning plant. So why not have a tiny power plant in each home?

- **A central power plant is efficient.** It can produce electricity much more efficiently compared to a million little home-based generators.

- **A central power plant is easier to maintain.** If you were running an engine-powered generator at your house 24/7, it would need maintenance all the time. You would have to change the oil on your generator once a week and replace it once a year, maybe more frequently. For comparison, people run a typical car engine one hour a day, not 24/7.

- **A central power plant is more reliable.** People need power at all times. If your home system goes down, your house becomes unlivable as everything in your refrigerator spoils and your home heating and air-conditioning system shuts down.

- **A central power plant can handle variability in power demand.** There will come a time in the average household when many appliances are running at once. To handle this combined load, the house needs a generator that can occasionally create, say, 20,000 watts of electricity. But most of the time, the house needs only one tenth of that, meaning that this big 20,000-watt engine-powered generator would be overkill most of the time, and underutilized.

- **A central power plant produces far less pollution than a million little generators.** This is especially true if those million little generators are idling at one tenth of their maximum capacity most of the time.

- **A central power plant makes it possible to scale up technology.** It makes sense to apply the highest levels of technology to a power plant serving a million people. For example, one house cannot have a nuclear power plant, while a nuclear power plant that serves a million houses makes sense because of the economies of scale. A $10 million scrubber system for a large coal-fired power plant makes sense if it only costs $10 per house. But there is no way to spend $10 and get any significant benefit for one house running on a generator.

As you can see, there have been important benefits to using big central power plants. Even today, when an individual house could theoretically generate its own power with solar panels, it still makes sense to be on the grid. The only way to run a house independently on solar panels is to have a big bank of batteries in your house to store the power so you have electricity at night and in rainy weather. Those batteries are still way too expensive to justify installing on a house-by-house basis. By connecting to the grid and using its power at night, those batteries are not necessary.

The problem with the grid, however, is that it can be vulnerable. The great Northeast blackout of August 14, 2003, shows how this vulnerability can express itself. August is one of the hottest months in the United States, so a lot of air conditioners were running and using electricity. Around 2 p.m., the start of the hottest part of the day, power plants were pumping out power near maximum capacity. The transmission lines were carrying power to customers near maximum capacity, too. Because the air in August is so hot, and the electricity running through the lines makes them hotter, the transmission lines started to expand and sag. One important transmission line sagged into a tree and shorted out. It went offline.

The power that that transmission line had been carrying still needed to get to customers, so it was diverted to the remaining transmission lines that were still working. Another transmission line failed from the overload. This diverted more power to the remaining lines. So another line failed, and then another. This is a classic example of a cascade problem, where the failure of one component makes things worse for other components, and under the extra load they all start failing. Fifty million people were without power for days; they did not have air-conditioning or refrigerators in the hottest month of the year. And nothing was actually broken—no one had cut the lines or lit any substations on fire.

It is not trivial to restart a power grid that has failed this badly, and it took several days to restore normal service. One big challenge is that when power is restored, everything that has been without electricity, from refrigerators to air conditioners, demands power all at once, again overloading the system and shutting it back down. To avoid this problem, little parts of the grid have to be brought up in sequence, and that takes time.

Now imagine that a group of coordinated terrorists bent on destruction and sabotage comes along and wants to cause a cascade failure by inflicting physical damage on the grid. With cutting torches and angle grinders, they cut down transmission towers. They shoot holes in big transformers at substations, so all the cooling oil drains out, causing the transformers to overheat and fail. They bomb substations with simple homemade bombs or pour gasoline on the control panels and light them on fire. Most of this grid equipment is completely out in the open, and much of it runs through unpopulated, rural areas with no one guarding it. It might take weeks, even months, to patch everything up so the grid can operate again. Meanwhile, a hundred million people are without power, and the economic engine of a third of the country grinds to a halt. The long-term effects of an attack like this would be uncomfortable, to say the least.

Another threat that the power grid faces is cyberattacks. These are online infiltrations that attempt to disrupt the computers that are increasingly used to control the grid (as well as other forms of infrastructure, like our financial system and the internet). Cyberattacks have been in the news recently for three reasons:

1 There are documented cyberattacks that now occur regularly. These attacks generally have to do with the theft of personal information. Famous examples include the Equifax data breach, where important identifying information for 145 million customers was stolen, and the Yahoo! data breach, where the hackers gained access to personal information for as many as 3 billion accounts.

2 Ransomware attacks have occurred on a number of businesses and municipalities with increasing frequency. In a ransomware attack, the attackers can remotely lock vulnerable computers and demand money to unlock them. For example, in 2019, one attack hit twenty-two Texas towns.

3 There have also been real-life cyberattacks on power grids, like a 2019 incident in the western United States that marked the first successful attack on a grid. This attack was minor and did no lasting damage, but it is significant because it was successful.

The challenge with cyberattacks is that they often exploit unknown security holes. The only reason that the holes are discovered is because the attack occurs.

PREVENTION

Many of the prevention steps described in the chapters on EMP attacks (page 44) and CME events (page 158) also apply to this doomsday scenario. Building overcapacity into the grid to avoid cascading failures, as described on page 9, would help, and the ability to operate important parts of society, such as gas stations and traffic lights, during a grid failure would be beneficial.

The smart grid, once fully deployed, could also help, as mentioned in the sidebar on the next page. The power grid also needs a better, more active security system. A great majority of the grid's equipment is sitting out in the open, and a lot of it (especially transmission lines) is in lonely places spread all over rural areas. Anyone could drive up to a transmission tower in a remote area and knock it down with a cutting torch. Substations face the same types of threats. Systems must be invented and deployed to detect these attacks and to prevent these kinds of attacks from occurring. For example, cameras and proximity sensors can monitor transmission towers and substations, alerting drones and guards to threats immediately. Grid security like this will increase the cost of electricity a little, but it will also help prevent a catastrophic attack on our grid from ever occurring. Right now, the grid's vulnerability is a huge security risk and contains a million points of failure that could be exploited to devastating effect.

There are also creative solutions that could improve the resiliency of individual homes and businesses. Battery systems are going down in cost and could make a house energy-secure for 12 or 24 hours if the grid goes down. In theory, any house with an electric car in the garage could be powered by the car's battery in an emergency if systems were designed to make this possible. New solid-state transformers could make older oil-cooled transformers obsolete and much easier and cheaper to replace. Substations could be housed inside secure and guarded buildings instead of being left out in the open. There are many possible solutions if the threat to the grid is taken seriously.

In today's modern world, electricity is nearly as important to any city, town, or household as oxygen is to a biological organism. Without power, a city is dead. Given its importance, the modern world should put a lot more emphasis on end-to-end grid security before someone takes advantage of the myriad security holes available today.

ABOVE High-voltage power transmission lines run through a field and past a railway line. Without security, transmission lines such as these can be vulnerable to sabotage.

SMART GRIDS

The latest development in power grids is called a smart grid. In the past, power grids have been passive systems consisting of wires and transformers. The only part of a traditional power grid that had to be "smart" was the power plant itself, as it had to adjust to changing power demand throughout the day.

With a smart grid, there is much more intelligence built into the system, all the way down to individual homes and businesses. If a transmission line on a smart grid fails, the grid can send a signal to all homes and businesses in the affected area telling them to shut off nonessential appliances like refrigerators and HVAC systems for thirty minutes. This greatly reduces power consumption, preventing a cascade of additional failures and allowing the transmission line to be restored.

BELOW One important feature of a smart grid is the ability to handle many different sources of power scattered across the grid.

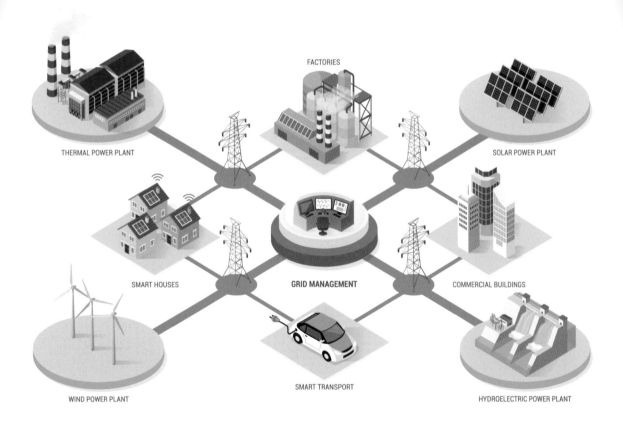

THERMAL POWER PLANT

FACTORIES

SOLAR POWER PLANT

SMART HOUSES

GRID MANAGEMENT

COMMERCIAL BUILDINGS

WIND POWER PLANT

SMART TRANSPORT

HYDROELECTRIC POWER PLANT

THE AUTOMATION ECONOMY

THREAT LEVEL: WORLD

"Impoverished" does not do justice to the squalor in which my husband and I lived in Bangladesh. I had been able to eke out a survivable existence for myself working in the ready-made garments (RMG) industry, but by Western standards I lived in appalling conditions. To earn a "living," I sewed together the clothing for consumers in North America and Europe, and in return I received slave wages. But at least I had wages, and I was therefore able to eat and sleep with a roof over my head.

If you don't live in a place like Bangladesh, Vietnam, or Indonesia, you simply cannot understand what daily life is like for millions upon millions of people in the developing world. But I lived this life every day. A garment worker in the RMG industry could theoretically make a minimum wage of 8,000 taka per month—the equivalent of $100 per month or $3 per day. But many factories did not pay even that. Just about anyone can understand that trying to live on $3 per day is absurd, but let's break down what this really means. I lived in Dhaka, in a squalid little space with my husband. In the United States, this tiny space might have been considered suitable as a stable for a farm animal like a goat. I spent 85 percent of my earnings to rent it per month. The remaining 15 percent, or about $13 for the month, needed to cover everything else. Since that is impossible, I was glad that my husband also had a job. With his money, we bought food, clothing, and other necessities. We cooked in a tiny open kitchen that everyone in the building shared. At the factory, I was given strict hourly quotas. If I missed them, even due to something like a machine

failure, I worked overtime for free. But all things considered, we had it "good" here. The conditions at a leather factory were even more dreadful.

No human being wants to work this way or to survive on this tiny sliver of money. Why was it happening, anyway? Because without a job, you starve. Any job that prevents starvation is better than nothing. Millions of people in Bangladesh depended on these jobs, as terrible as they were, for survival, and it was a race to the bottom with wages under these circumstances.

The wages that we earned were an atrocity. The brands selling these clothes could have just as easily paid us 500 taka (about $5.85) an hour instead. The shirts or pairs of jeans would have cost a dollar more in America, but so what? It is not like the people in America or Europe could not afford that. The reason they paid us so little was because our

RIGHT Industrial embroidery machines have allowed manufacturers to automate a task that humans once handled. Similar advances in automated sewing machines will eventually replace the human workers who sew cloth into clothing.

desperation meant that they could. Desperate people in unregulated, impoverished countries work for pennies because they have no other choice.

But now, with automation on the rise, even these slave wages would be welcome relief to hundreds of millions of people all across the developing world. As automation steals more and more jobs, the people who need the work have fewer and fewer options.

An economic crisis occurred in Bangladesh when sewing robots arrived. At first these robots could only do very simple, very specific sewing tasks, but the new models kept improving and eventually could piece together any article of clothing. Soon, the garment industry in Bangladesh did not need any people at all to operate sewing machines. Since factories filled with automated sewing machines could be located anywhere, thousands of companies moved out of Bangladesh. The economic crash that followed was swift and deadly. Millions of people in Bangladesh lost their jobs, terrible jobs though they had been.

The same thing happened in the shoe industry. Clever engineers developed automated machines that could assemble a pair of shoes without any need for human intervention. Then the electronics-assembly industry followed suit. Millions and millions more people in developing countries lost their jobs.

Automation reached North America, Europe, Japan, and China. All the car factories became fully automated, as did the fast-food restaurants and retail stores. Truck drivers, taxi drivers, and ride-share drivers everywhere lost their jobs to autonomous vehicles. Many construction jobs were automated as well. Countless people, especially in the working class, lost their livelihoods around the world as automation quickly made all their jobs irrelevant. Unemployment and poverty rates shot up. The United States, which had already been showing economic strain before the automation wave, turned into a desperate and poverty-stricken place for millions more.

Our current capitalistic economic system has no solution. Under this system, there is no obligation for any company to employ people. Human workers are expensive and troublesome. As soon as automation offered a way to eliminate human workers, companies jumped at the opportunity. Those who lost their jobs, whether it was the tens of millions of working-class people in the United States, the millions of garment workers in Bangladesh, or the armies of people assembling the pocket electronics for tech companies, were all doomed when their jobs evaporated to automation. I was forced to return to the impoverished rural village of my parents, where we eke out a living growing food on our small plot without access to health care, clean water, or sanitation. It was awful to return to the terrible housing and hunger, but there was no other choice.

SCENARIO

Many of the doomsday scenarios in this book have economic effects. But is it possible for an economic crisis to cause a doomsday scenario? If we define a doomsday scenario as something that impoverishes the lives of millions of people or ruins an entire country's economy, then the answer is certainly yes.

We've seen one example with the Great Depression. The Depression started in the United

ABOVE Unemployed men line up outside a shop offering free food in this Depression-era photo.

States after a stock market crash in 1929, and it lasted through the 1930s. It was a doomsday situation for millions of people. The US unemployment rate rose as high as 25 percent. Half a million economic refugees known as "Okies" (because many of them were from impoverished states like Oklahoma) were simultaneously evicted from their midwestern homes and land and forced to move. This terrible and unfair migration is depicted in John Steinbeck's book *The Grapes of Wrath*.

Likewise, the economic collapse of Venezuela that accelerated in 2015 due to a drop in oil prices ruined the lives of the country's 28 million citizens. In 2019, it was estimated that the percentage of Venezuelans who were living in poverty increased to 94 percent. One study found that the average Venezuelan lost 19 pounds (8.7 kg) in body weight in 2017 due to poor nutrition and food insecurity—a phenomenon nicknamed the "Maduro Diet" after Venezuelan president Nicolás Maduro. In a starvation situation like this, the effect on growing children can be especially dire because it stunts their growth as well as their mental development.

There is an oncoming economic crisis that has the potential to create a doomsday event like this on a global scale: one caused by the approaching automation wave.

Think about the jobs that you see around you in your day-to-day life. You go to a grocery store. People stock those shelves, clean the floors between the shelves, and check people out at the cash registers. Grocery stores are largely unchanged over the last century in terms of the people working there. When you eat at a fast-food restaurant, you'll find people cooking the food, taking orders, and cleaning the tables. At a construction site for a new house, human workers build the frame for the house, put on the shingles and the siding, install the drywall, and paint. All this work has been completed with human labor for centuries.

As we will see in the Science section, new technology will be eliminating jobs like these and many others around the world. Hundreds of millions of jobs are at risk worldwide, especially in developing countries where manual labor at slave wages is a situation that is considered "better than starving." We get incrementally closer to this outcome every single day. Why? Because each bit of automation we create, each new robot we deploy, makes products cheaper and easier to produce. If there are two companies doing the same thing, and one of the two uses more automation, then the automated product will be cheaper, and many customers will be inextricably drawn to the lower price made possible by automation.

We can see how automation can take over by looking at farming and textile production over the last few centuries. If we were to go back to America at the time of the Revolutionary War, farming was almost completely human-powered, with the occasional draft animal thrown in. Let's say that in 1776 we wanted to grow cotton and make a shirt from it. We had to prepare the soil, using a shovel or with a plow pulled by a horse or ox. We could plow perhaps two acres a day. To plant the seeds, we walked through the fields and planted them one by one. We would weed the crop by hand with a hoe. When it was time to harvest, we would walk through the field with a sack and collect the cotton bolls one by one. We would bring the sack of bolls back to the house to pick the seeds out of it by hand. Then we would card the cotton, spend hours spinning it into thread with our spinning wheel, and then put the thread on our hand loom to spend many more hours to make cloth. Finally, we would cut the cloth and sew it together with a needle and thread. It took a gigantic amount of human labor to do anything agricultural. To make a T-shirt in 1776, we would need hundreds of hours of human labor when it was all said and done.

Today, instead of hundreds of hours, a T-shirt needs just a few minutes of human labor. That's because a giant cotton farm today with thousands of acres can be operated by a tiny group of people and a lot of automated farm equipment. Soon, it may not require any people at all except for an overseer and an occasional maintenance worker, as robot tractors are almost here. Automated machines now gin, spin, and weave the cotton into beautiful flawless cloth. The very last step still requires a person to operate a sewing machine to assemble the T-shirt, but this only takes minutes of human labor. Sewing the shirt together is the only labor-intensive step left. And this last step is also about to be automated. In the not-too-distant future, we will be able to make a T-shirt without any human labor at all.

Meanwhile, the upper-level positions at large companies that depend on this manual labor are secure. While the workers who assemble shoes, clothing, and electronics in places such as Indonesia, Vietnam, and Bangladesh receive incredibly low wages—on the order of $200 per month, or less than $1 per hour—for their work, executives are paid millions and millions of dollars on average. With automation, executives will likely continue to make more and more, as has been happening consistently over the last fifty years.

The world by and large has hitched its wagons to a capitalistic economic system that often serves

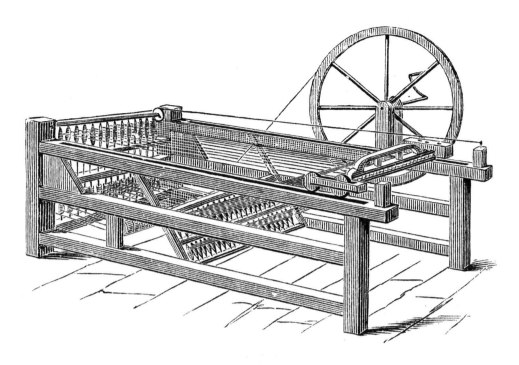

ABOVE The spinning jenny, invented in the mid-1760s by James Hargreaves, helped usher in the Industrial Revolution by making it much more efficient to spin thread and yarn for weaving.

the interests of business owners and the wealthy at the expense of everyone else. This system does not always offer strong incentives to help workers keep their livelihoods, offer living wages, or create good jobs, an effect easily understood by looking at Bangladesh and the many countries that share its fate. Today, 70 percent of the people on Earth make $10 per day or less. In the United States, money flowing to the wealthy has exploded, increasing by 420 percent since 1980, while the income of rank-and-file workers has barely grown since 1980. This same phenomenon is happening around the world.

With the approaching wave of automation, some of capitalism's worst features are about to manifest themselves. The rise of robots in the workplace should in theory be beneficial for everyone on the planet, because as robots do more and more work, humans should need to do less and less. Unfortunately, with the way that capitalism is wired

today, automation can make life far worse for the millions of people who will lose their jobs.

If this trend in inequality continues, automation will lead to an increasing concentration of wealth among the top 0.1 percent of people in the world, along with stagnating or decreasing middle-class and working-class wages. To see a very simple example of what might happen, let's consider the approximately 4 million people in Bangladesh operating sewing machines right now for $100 per month. Once sewing machines become vision-capable and fully automated, these 4 million jobs will be lost. No new jobs will be coming along to replace the jobs that are lost, and automation will be replacing many other working-class jobs across the board. Therefore, these 4 million workers in Bangladesh will become unemployed, and may never be employed again, unless radical changes occur in the world economy.

THE DOOMSDAY BOOK

In the car factories around the world, robots now complete a good portion of the work. Robots handle all the welding and the painting of a car body, but when it comes time to put in the wiring, seats, and carpet, there are still lots of people doing this work in most factories.

Why? Why was the welding easy to automate fifty years ago, but other parts of car production still need people? Why do we still have people sewing clothing and shoes, assembling smartphones, or putting hamburgers together in a fast-food restaurant? It largely has to do with the absence of good computer vision systems.

Welding can be fully and easily automated because a welding robot does not need to be able to see. Precisely formed pieces of metal are loaded into very precise jigs, and the welding head can perform the weld by dead reckoning. In other words, the metal in the jig cannot move, and the robot knows its exact shape, so the robot can repeat the exact same pre-programmed weld path time after time with high precision and zero defects. No vision is necessary.

Compare that task to restocking a shelf in a grocery store. This restocking task cannot be done without a vision system because it is unknown how many items are left on the shelf at any given time, or whether they are in the right places, or how badly the pesky customers visiting the store have messed things up. What if a customer is looking at shampoo in the shampoo aisle, decides he really does not need the hot dogs he picked up five minutes earlier, and leaves them on the shampoo shelf? As infuriating as this behavior is, it happens all the time. A human

LEFT In 2019, Ford showcased a prototype of a delivery robot designed to work with a self-driving car. When the car arrives at its destination, the robot unfolds itself from its companion vehicle and carries the package to the recipient's doorstep   BELOW A robot at a car manufacturing plant completes a welding job.

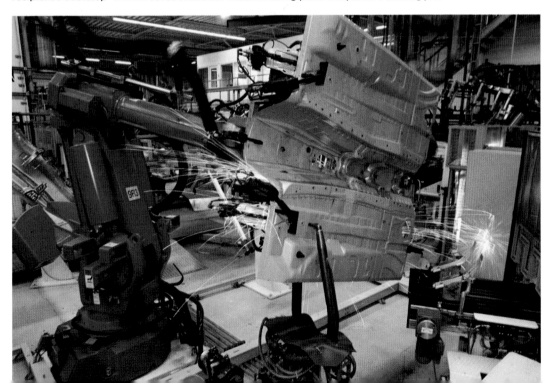

worker can look at the shelf, remove the misplaced hot dogs, rearrange things neatly, and then put new items on the shelf where they belong. Until a robot can do that same kind of vision processing, robots cannot restock shelves, pick tomatoes, sew a shirt together, and so on.

Thus, there have been hundreds of millions of jobs on the planet that have been protected from automation by this lack of computer vision systems. And the unfortunate problem for the people holding these jobs is that the vision systems to replace them are about to arrive. Engineers have finally improved silicon chips to the point where their power starts to rival that of the human vision system. Meanwhile, software engineers have been working on the algorithms that let cameras see the world like people do. And once these visual capabilities are married to precise and flexible robotic hands, robots will be able to take over millions of the remaining unautomated working-class jobs in restaurants, retail stores, assembly lines, construction sites, garment factories, and more. All the jobs that have traditionally been protected because robots cannot see will suddenly evaporate.

How will computer hardware engineers and computer scientists replicate the capabilities of the human vision system? The first thing to understand is that general vision processing can take an impressive amount of processing power.

In the human brain, the occipital lobe handles vision processing. The occipital lobe makes up about 18 percent of the human brain. In very rough terms, if we consider the brain to be equivalent to a computer that can process 40 quadrillion operations per second (40 petaops), then the occipital lobe is running on the order of 7 quadrillion operations per second. At this moment in history, this is still considered supercomputer levels of computing power, where the supercomputers are room-size machines that take a lot of electricity and HVAC capacity. Humans still have a long way to go before completely replicating the power of the occipital lobe in a laptop-size device, but we are getting within striking range. Intel now has individual AI chips that can do roughly 120 trillion operations per second. This is about 1.5 percent of the occipital lobe's computation power. For very specific and well-defined visual processing tasks, it is more than enough to get the job done.

What can AI specialists do with a chip like this? They can:

- Recognize faces of individual people with good accuracy

- Learn what thousands of objects look like and then recognize them on a shelf

- Identify items on a conveyor belt or table and pick them up, find defects, fix things that are out of place, and perform other quality control tasks

- Look at medical samples (e.g., under a microscope), find cancer cells or diseased cells with better accuracy than humans, and review X-rays and other outputs from medical imaging devices like CAT and MRI machines

- Monitor security cameras and raise an alarm when something is wrong

Blue-collar jobs are not the only thing robots will threaten. White-collar jobs are also in danger because of advanced artificial intelligence and economic pressure.

Think about doctors. Doctors are extremely expensive to educate and train, and there is a growing demand for their services. At the same time, the complexities of the job are mounting. It is nearly impossible for a human doctor today to know everything needed to provide good holistic care and wellness for one human patient, never mind the typical caseload a doctor sees on an annual basis. A doctor needs to keep track of all the new medical research and best practices coming out each day, new drugs and devices, new treatment and procedure options, and all the potential drug interactions and drug differences between different body types. Medical errors from human doctors kill hundreds of thousands of people a year in the United States alone. Given all these factors, there is huge and growing pressure to replace doctors with much more capable AI physicians and surgeons. Robotic arms are already performing many surgeries with human guidance. It may only be a matter of time before the robots take over the whole task and do it better than humans.

Similarly, in court cases, there is plenty of evidence that many judges and juries harbor unconscious, implicit biases, and that judges rule differently depending on whether the ruling comes before or after

a meal. Replacing fallible humans in these roles with AI would create a much fairer and lower-cost system.

The technology that will power these white-collar developments was first embodied in Watson, the unbeatable *Jeopardy!*-playing robot from 2011. The technology behind Watson allowed the algorithm to digest gigantic amounts of free-text information, including, for example, the entirety of Wikipedia, IMDB, and an enormous dictionary. Watson could answer questions using this knowledge. Today, neural networks are enhancing these results, so that AI algorithms can now do remarkably well on human IQ tests, and AI chatbots can converse with human beings convincingly on almost any topic. As this technology is refined, AI doctors, financial advisers, pilots, nurses, lawyers, and chemists are inevitable, and they will be far better than the human equivalents whom they replace.

BELOW Robots can assist in a wide range of surgical procedures, including certain types of tumor removal and hernia repairs.

PREVENTION

It is easy to understand where this wave of automation could take us if we look at its end point. Eventually, we will be able to create a civilization where robots do all the work that humans have traditionally performed. Robots will grow, distribute, cook, and present all the food humans eat. Robots will make and ship all the clothing we wear. Robots will build and maintain all the housing everyone needs. Robots will treat all our medical conditions and heal us. Robots will educate and entertain us. Robots will make all our cars and airplanes, smartphones, furniture, appliances and . . . well, everything. There will not be a single thing left that humans need to do. In theory, when we arrive at this point, it should be glorious. With robots doing all the work, every human should be able to live in luxury on perpetual vacation.

But we can look at the economy today and see that we are not headed in this direction at all, and it is already causing significant pain. As new robotic vision technology causes the next big wave of automation, it will be an economic doomsday for millions and millions and millions of workers unless we change the rules of the economy to account for robots.

So how do we prevent this doomsday scenario from happening? The key is to understand the eventual end point—an economy completely manned by robots—and redesign the economy so that every human on the planet *can* actually live in luxury on perpetual vacation. We start by modifying the system we have today to make it more equitable while transitioning ourselves to the new economy. Let's explore this approach for the moment.

We need to recognize that society gains nothing from the inequality and concentration of wealth that we see happening today. Nor from the poverty seen around the world today. An economic system

BELOW Automation will eliminate millions of jobs and increase unemployment, making it harder for low-wage workers to afford basic necessities, such as high-quality housing.

that gives everyone on the planet access to necessities, including good food, clean water, safe housing, comprehensive health care, and effective education, is essential. And this must also happen in a context that reverses all the damage to the environment that imperils the global ecosystem and thus the survival of millions of species. In other words, we as a society need to consider the greater good, and even more so the highest good. Simply ask ourselves: What can the new robotic economy maximally accomplish for everyone on the planet, and for the planet itself?

The first thing to do is to completely eliminate the concept of a "billionaire" from this planet, and in fact go much further, perhaps capping maximum wealth at $5 or $10 million. This is the kind of position staked out by the French economist Thomas Piketty. He proposes that we do this with wealth taxes. In addition, put strong inheritance taxes in place and enforce them. For this plan to be effective, we would need to work together across the planet to eliminate all the tax shelters that wealthy individuals and corporations now use to hide money and assets from taxation. With extreme inequality and the concentration of wealth eliminated, and basic needs accounted for, two things can happen: prices plummet as automation replaces all human labor, and the money raised can be distributed to everyone else.

The second thing to do is to implement and strengthen systems that provide basic needs. All developed nations but one today provide some form of universal health care for their citizens. We can make health care universal across the planet, with everyone receiving the same level of excellent care, just as every developed nation provides public education. The same philosophy can apply to housing, food, and other necessities. In Singapore, 80 percent of the population lives in high-quality housing provided through the government—we can learn from

efforts like these and expand them to a global scale.

How do we eliminate this horrific statistic, that 70 percent of people on the planet make less than $10 per day, along with all the privations at this level of poverty? One way would be to set a minimum standard of living that provides everyone on the planet with great housing, excellent food, fashionable clothing, modern and ubiquitous health care, and education, allowing robots and automation to provide it all. Universal basic income (UBI) can fund this. The concept is promoted by a wide variety of luminaries, including Tesla founder Elon Musk, venture-capitalist investor Tim Draper, economist Milton Friedman, World Wide Web inventor Tim Berners-Lee, and more. Since the end point of automation is that robots perform all work, two results are possible: either all of their productivity flows upward to a wealthy few (as is happening now), or we distribute the gains from this productivity to everyone (as it should be). UBI is simply a mechanism to distribute the abundance and wealth that robots produce to everyone evenly.

We as a species have never risen to the level of considering the needs of all human beings and of designing a global economic system that can provide for those needs. If humanity were able to make this mental transition while at the same time taking into account the needs of the environment, things would be immeasurably better for everyone on the planet. Our failure to make this cognitive leap causes immeasurable suffering for billions of people.

The key is to understand that the coming wave of automation will be devastating to millions of workers unless we band together and create an economy that works for the highest good. Rather than leaving billions of people impoverished, a new economic system will be able to deliver all of the benefits of automation to everyone on the planet evenly, as well as saving the planet itself.

OPIOID CRISIS

THREAT LEVEL: COUNTRY

You know how there are certain days in your life that are burned indelibly into your memory? For me, the first time I took oxycodone was one of those days.

I had gone to the dentist first thing in the morning to have my four wisdom teeth extracted. My husband went with me to drive me home afterward. The procedure went smoothly, and I got "a prescription to help with the pain." My husband didn't give the prescription a second thought. He filled it at the pharmacy and handed me the bottle of pills around lunchtime that day. The instructions on the label said to take one pill every twelve hours as needed for pain.

When I took the first pill, the effect amazed me. It was like this incredible weight had been lifted from my shoulders—anxiety and worry and dread that had been a part of my life forever disappeared when I took that pill. For the whole day, I experienced a feeling of warmth—a glow—and the sense that everything was going to be okay. The pill helped with the pain of the missing wisdom teeth, sure, but that was a very minor benefit compared to the relief the pill offered for the pain of life. I could not believe how incredibly freeing it was for that pain to be gone. If you had lived your entire life dragged down by a constant undertow of anxiety and worry and fear, can you imagine how good you would feel if it all went away?

When I woke up the next day, the pill had worn off. All the anxiety and dread was back. My missing teeth felt pretty good—aspirin could have handled any remaining pain. But I wanted another oxycodone pill. I wanted that feeling again where all of the stress and worry were completely gone from my life.

At the time, I didn't know oxycodone was an opioid, like heroin. It was a medicine that my dentist had prescribed. Because he told me to take it, I took it—innocently, blindly. They were pills, and I took them as instructed.

Like all prescriptions, the bottle emptied, and I faced the bottom of the bottle with dread. This situation was nothing like running out of antibiotics or blood-pressure medication. I did not want this feeling to end. I needed these pills. I called the dentist's office, and they obliged by refilling my prescription. But something else was happening. The pills didn't work as well anymore. I found that I now needed to take two at a time to get the same effect.

When I ran out again, I called the dentist's office. This time, they were worried that I had used up the prescription too fast. They also felt that I should not be suffering from pain anymore. While they were polite about it, they refused to refill the prescription. I could come in if I wanted, and the dentist could look at my teeth. Since I knew my teeth were fine, I made an excuse, said I would call back, and quickly hung up the phone.

RIGHT Humans have used opioids in different forms for centuries. This photo shows modern opioids: a bottle of oxycodone pills, hydrocodone tablets, a syringe for heroin, and fentanyl powder.

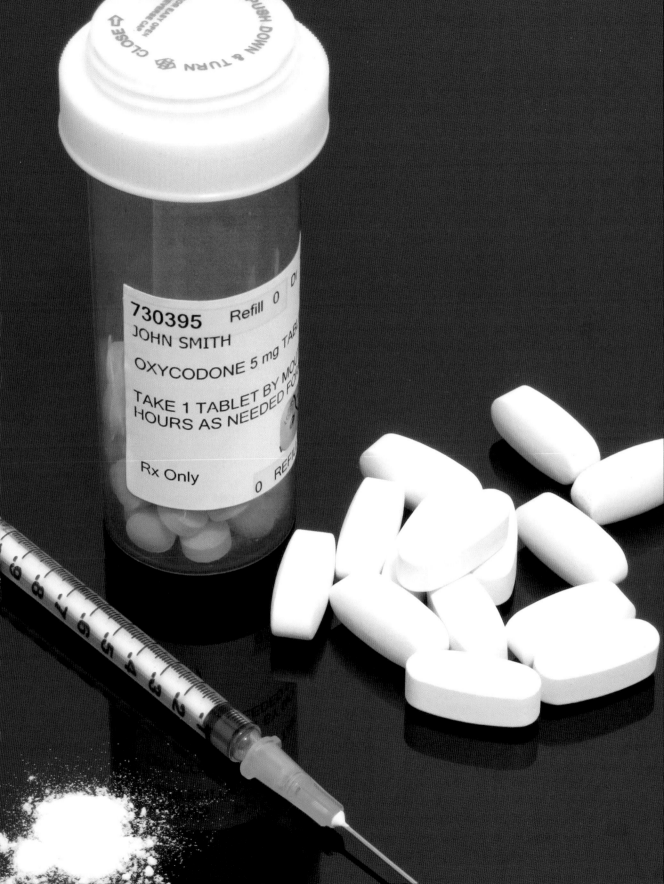

Now what? As my last pill wore off, I started to feel awful as all the old feelings came back. Then another level of awfulness began. Later, I would learn that this was my first taste of physical withdrawal. Imagine the flu, but ten times worse. I knew I *had* to get more pills. I knew my mother had some in her medicine cabinet, left over from her back surgery last year. I went to her house, made an excuse to use the master bathroom, and stole half her pills. When I looked at the label, I was glad to see that hers were a stronger dose than the ones I had been taking; I was glad, because I needed a stronger dose.

I have no idea what would have happened if I had continued down the road I was on. Fortunately for me, reality intervened. Kaylee's death was a real wake-up call. Kaylee had been a good friend in college, and we had both moved to the same city when we graduated. We'd stayed in touch. But I had not heard from her for several months. Her husband Jim called to give me the news about her death.

Kaylee had overdosed on heroin. She had gotten into pain pills the same way I had—her doctor had prescribed them after surgery on her ACL. She had started using heroin when she ran out of pills because it is a lot less expensive than pain pills on the black market. But one batch of the heroin she'd bought was laced with fentanyl. Kaylee overdosed and died in their bedroom from asphyxiation. Jim found her dead on their bed when he got home from work.

I could clearly see where I was headed, and it was enough to wake me out of my haze. In tears, I confessed the whole thing to Jim right then and there. He agreed to go with me to talk to my husband, and they got me into rehab. I have been in recovery now for a year.

I still think about opioids every single day and the incredible feeling they gave me. I know that my only hope is to stay in recovery, even though it feels terrible, and I pray that I will not backslide. But sometimes it is so hard. . . .

SCENARIO

Is America's opioid disaster a doomsday scenario? We often think of a doomsday event as quick and catastrophic, like an asteroid strike (see page 122), a nuclear bomb (see page 22), or a tsunami (see page 150). While America's opioid crisis has unfolded in slow motion starting in the late 1990s, killing over a half million people in the United States, it has certainly been catastrophic enough to be considered one.

The death toll is sobering. More than 50,000 people per year now die from opioid overdoses. Many people who began misusing opioids were simply following and trusting their doctor's instructions, as they would with any prescription medication.

About 25 percent of the people with chronic pain who receive prescription opioids will end up misusing the pills. Some of these patients switch to heroin either because they cannot get pills or because heroin is less expensive. Once people are addicted to heroin, the probability of overdosing increases. Prescription painkillers are a gateway drug for heroin, with 80 percent of today's heroin users starting out with prescription painkillers.

People of many different backgrounds and ages, from middle-age parents to teens, became addicted and died from overdoses after starting prescription painkillers. There are hundreds of thousands more of

TYPES OF OPIOIDS

- **Opium:** opioid made using the natural sap from the opium poppy
- **Morphine:** natural opioid extracted from the sap of the poppy, ten times more powerful than opium
- **Codeine:** a narcotic also extracted from the sap, less potent than morphine
- **Heroin:** a chemically modified version of morphine that is twice as powerful
- **Tramadol:** a synthetic opioid that is one-tenth as potent as morphine
- **Hydrocodone:** a synthetic opioid as strong as morphine. The prescription painkiller Vicodin® combines hydrocodone and acetaminophen.
- **Oxycodone:** a synthetic opioid as strong as heroin. Oxycontin® is a time-release version of oxycodone. Percocet® combines oxycodone and acetaminophen.
- **Fentanyl:** a synthetic opioid that is 100 times more powerful than morphine
- **Carfentanil:** a synthetic opioid that is 10,000 times more powerful than morphine

these stories where lives and families have been ruined. An uptick in opioid-related deaths is happening in other countries, including Australia and the UK.

This is a doomsday scenario not just in terms of its staggering death toll but also in terms of its huge economic cost. In the United States, it is estimated that half a trillion dollars per year is destroyed due to lost productivity, healthcare expenses, and other costs. In addition, most of the heroin and fentanyl coming into the United States comes from foreign countries. In essence, foreign entities are killing tens of thousands of US citizens per year, resulting in what resemble wartime death statistics. In this sense, drug trafficking starts to look like a national security issue.

What went wrong? How did this crisis get started? Why is it legal to sell a drug that leads to the deaths of so many people? Why would the company keep selling a product that is obviously killing people? Why would the government and the medical establishment allow the company to keep selling the product? Why

would doctors prescribe opioids, which are known to be highly addictive and destructive, to their regular patients in record numbers? And why were people receiving powerful opioids for routine health concerns like kidney stones, short-term pain from surgery, and wisdom teeth extractions in the first place?

In *Missouri Medicine*, Dr. Ronald Hirsch points out several different factors that allowed this doomsday scenario to unfold in the United States:

- Purdue Pharma started running wall-to-wall ads in medical journals and distributing marketing materials directly to doctors, touting the benefits of Oxycontin, which the FDA approved in 1995. The primary selling point of the drug was its time-release formulation, which was said to provide twelve hours of pain relief. These ads influenced doctors but underplayed the risk of addiction.
- The FDA approved Oxycontin based largely on

the company's clinical trials and other testing, but some of the results were spun to create a narrative that the drug was more effective than it was. The clinical trials did not show any important efficacy improvements and also downplayed addiction risks. The fact that the FDA gave its blessing would certainly influence doctors.

- The American Pain Society had just launched a campaign calling pain "the fifth vital sign," ranking it with blood pressure and temperature as important indicators for doctors to measure and track. Unfortunately, there is no way to objectively measure pain, and pain is also easy to fake.

- The Joint Commission, an organization that "accredits and certifies over 22,000 healthcare organizations and programs in the United States," also began emphasizing pain management. It published a guide for pain relief promoting the new opioids. These campaigns, too, would influence doctors, leading them to believe that opioids were safe for pain relief.

- Press Ganey, an organization that gives patients surveys and questionnaires about their experiences at hospitals, prioritizes "patient satisfaction" as an important

indicator of a hospital's performance. The Centers for Medicare and Medicaid Services (CMS) also started to emphasize patient satisfaction. This creates a feedback loop that would influence doctors who prescribe drugs. If a patient asks for opioids and the doctor refuses, the patient could punish the doctor by giving a low score.

All these factors acting together encouraged doctors to freely prescribe Oxycontin and similar drugs despite potential problems. Between 2006 and 2012, 76 billion opioid pills were prescribed in the United States. That's an average of 11 billion per year. For comparison, Americans only consume 29 billion aspirin pills per year. And aspirin is a widely available over-the-counter medicine that costs pennies per pill. In one small town in West Virginia, between 2001 and 2012, tens of millions of pills were distributed through the town's pharmacies even though the town's population was only 3,000.

If this doomsday scenario, which killed hundreds of thousands of people and was driven in part by greed and corruption, happened once, could it happen again? Given the slow response and the growing death toll of the opioid epidemic, it seems quite possible.

SCIENCE

Opioid use goes back thousands of years and started with *Papaver somniferum* plants, a species known as the opium poppy. After its flowers bloom, a spherical seed head forms.

If you score a green and developing seed head with a razor blade, it will ooze a sticky sap for a few hours. When collected and dried, this sap becomes

opium. Historically, humans typically ingest opium by mixing it with water and drinking it as a tea, or by mixing it with tobacco and smoking it. In 1822, the autobiographical book *Confessions of an English Opium Eater* by Thomas De Quincey made people in the Western world aware of opium's euphoric effect and also the dangers of addiction.

Opium contains several different drugs, including morphine and codeine, and is also physically bulky. To shrink opium, a chemist can process the opium to extract the morphine. He or she can then alter the morphine to create a white or brown powder that we know as heroin, a process discovered in 1874. The heroin molecule is a slightly modified version of the morphine molecule that makes it more potent. Starting in 1898, Bayer trademarked the name "heroin" and sold the drug as a cough suppressant for more than a decade.

As just stated, heroin comes from morphine and therefore has its roots in the sap of an opium poppy. In that sense, it is "natural." A "synthetic opioid" like fentanyl, on the other hand, is entirely man-made and has no connection to opium. Fentanyl was invented in the 1960s and is far more powerful.

Molecules such as morphine, heroin, oxycodone, and fentanyl are all known as opioids. In the human body, these molecules dock onto mu opioid receptors in the central nervous system. These receptors exist inside our bodies because our bodies (specifically the pituitary gland) produce chemicals called endorphins, which reduce pain. An endorphin molecule binds to the mu opioid receptor in the brain and blocks a pain signal. The feeling known as a "runner's high" is attributed to the body's natural production of endorphins after physical exertion. It is thought that endorphins block pain while muscles recuperate. Morphine molecules (and other opioids) fit right onto mu opioid receptors as well, which is why they are used as painkillers.

Most opioid molecules also act on dopamine cell bodies in the "pleasure center" of the brain, also known as the nucleus accumbens. When activated, these cells release dopamine, which creates feelings of pleasure. In sufficient quantities, most opioids will induce euphoria. If many of these cell bodies can be induced to fire together, say because a person injects heroin, it creates intense feelings of pleasure.

LEFT Sap seeps through cuts made into an opium poppy's seedpod.
ABOVE Mrs. Winslow's Soothing Syrup was a product that was sold as a remedy for teething pain in children. It is an example of the many commercially produced medicines from the nineteenth century that contained morphine.

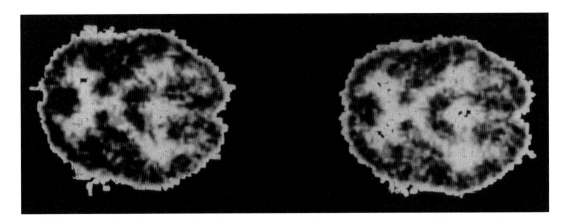

ABOVE These PET scans show the brain activity of a person with a substance use disorder. The image on the left shows decreased activity caused by the use of morphine, while the image on the right shows a more active brain without the influence of morphine.

These euphoric feelings will only last for a little while. If a person keeps taking opioids, the human brain quickly adapts and develops a tolerance to them. Now the person has to take more of the drug to get the same pain-relieving or euphoric effect. This is a common process that can also happen with drugs like caffeine, alcohol, and nicotine.

Meanwhile, the body does not build up tolerance to the other effects of opioids, like respiratory depression, nearly as quickly. This is one way that people die from opioids. In order to get the pain relief or euphoria that they want, people need to take more and more opioids over time because of the body's tolerance. But if they take too much, they stop breathing and die. Those who use street drugs, which can have very different purities or may be cut with ingredients like fentanyl, are at additional risk, because they do not know exactly what they are working with when they take a dose. It makes it even easier for those with an opioid addiction to miscalculate and take too much of the drug accidentally. These twin avenues to overdose are what make opioids deadly.

Unfortunately, if a doctor prescribes opioids and the patient is unaware of these problems, some

WHAT CAUSES OPIOID WITHDRAWAL?

If an opioid user stops taking opioids, the process of withdrawal begins within twelve hours or so. Withdrawal can happen with alcohol and caffeine, but opioid withdrawal can be especially uncomfortable. Its symptoms include diarrhea, vomiting, tremors, muscle pain, and sweating. The symptoms are triggered by an area of the brain known as the locus coeruleus, which contains a particularly high proportion of opioid receptors and reacts when opioid levels in the blood decline.

patients can easily get addicted. Once addicted, it is very easy for the patient to spiral out of control, as tolerance sets in and failure to take the drug causes withdrawal. Once the doctor cuts off the supply of prescription opioids, many of those who have become addicted switch over to heroin from street dealers, and the potential for overdose and death skyrockets.

PREVENTION

The idea of doctors and dentists routinely prescribing a drug that can easily ruin a person's life for something as mundane as a tooth extraction seems misguided and dangerous. But this has happened millions of times in the United States, and hundreds of thousands of people have died as a side effect of this process. How could we stop the current opioid epidemic that is killing 50,000 people a year and prevent similar drug epidemics in the future?

One effective strategy is to stop prescribing opioids, except for severe health conditions like cancer pain, debilitating chronic pain, and end-of-life palliative care. Meanwhile, the country could radically increase the research dollars being spent on pain-relief research, in an effort to discover less-addictive alternatives that can replace opioids.

The US opioid epidemic was also fueled in large part by marketing campaigns as well as false marketing from drug companies. One possible solution to problems like these is the concept of a corporate death penalty, as well as punishment for corporate executives, for unethical business practices. If a corporation causes the deaths of thousands of people, the corporation itself should be "put to death," its executives should be severely punished with fines and jail time, and profits should be recovered by clawing them back from recipients. Severe penalties like these would act as a deterrent.

For those who already have an opioid addiction, another way to help would be to decriminalize heroin. Portugal was able to do this successfully, starting in 2001, and has seen multiple benefits. Many of the problems associated with black-market and street drugs decreased, and effective treatment solutions for individuals with substance use disorders were an integral part of the program. Views on heroin use shifted as well. Heroin use is now seen as a health problem, and there is more emphasis on harm reduction for drug users.

If heroin were legalized, users would be able to take a reliable, pure product in controlled settings, vastly reducing the risk of death. It would also eliminate black-market heroin and all the associated risks for users. The price of opioids would also decrease. A dose of some synthetic opioids could be nearly free, because they are so concentrated that only a small amount is needed.

By combining the decriminalization or legalization of heroin with a strong network of treatment centers, help would be easily available for those who are ready to end their substance use. In addition, education efforts can help keep as many people as possible away from opioids in the first place. It would not be a perfect system—ideally, no one would be abusing opioids in the first place—but it would be better than a status quo where so many people are dying in the primes of their lives.

PART II
NATURAL DISASTERS

ASTEROID STRIKE

THREAT LEVEL: WORLD

We are standing on a beach in southern Florida. Millions of us. Along with billions of others around the world, we are looking up into the night sky. We see the moon, full and bright. They say the asteroid that is barreling toward Earth will just miss the moon, a cruel irony. With a slight difference in trajectory, the asteroid could have harmlessly slammed into the far side of the moon instead of Earth. Sadly, that's not going to be the case today.

Instead, an asteroid we cannot yet see is only six hours away from impact. It is flying past the moon as we watch the sky, but there is nothing to see now, even with the asteroid so close. In truth, there has been nothing to see with the naked eye for months, not since they originally announced this almost-incomprehensible life-obliterating catastrophe. The asteroid is said to be approximately the size of the one that killed the dinosaurs 66 million years ago, measuring 25 miles (40 km) in diameter and traveling at 40,000 miles per hour (64,000 km/h). It is headed on a direct collision course with the planet. It will not "skim" Earth. There will be no glancing blow. It is coming straight at the planet, like a bullet aimed at the center of a target. Impact is predicted to be somewhere just north of Cuba.

The official word is that there is nothing humanity can do to stop this asteroid. It was not detected early enough; humanity could not develop a response in time. We have been told that for people in North and South America, the end will come quickly. Millions of people in the United States have taken this prediction to heart and decided to short-circuit

the process. We have all traveled to southern Florida for a mass die-in.

The thought is that the explosion from the asteroid's impact will easily and instantly destroy most of the state of Florida. Therefore, for these millions of people, the end will be quick and painless. Others have driven as far north as they can go, in hopes that the predictions are wrong or overstated and there might be a chance of survival. Still others have traveled to the opposite side of the planet, to places like China and Australia, hoping for the best. Many wealthy people have gone a step further and are hunkering down in deep underground bunkers stocked with enough food and water to last for at least two years, hoping to survive the coming storm. Their goal is to stay underground long enough for the atmosphere to clear so that they can then emerge to rebuild society.

PREVIOUS SPREAD The Kobe Earthquake in 1995 caused some elevated portions of the Hanshin Expressway to collapse completely. RIGHT This concept drawing shows the impact of a massive asteroid. Geologists have found evidence that asteroid strikes have triggered immense earthquakes and tsunamis throughout Earth's history.

But the fact is, no one really knows what is going to happen. There is no shortage of high-tech simulations, both rational and rabid speculation, hyperbolic predictions, referrals to historical events, and, of course, religious allegories to the End Times, Armageddon, and the Rapture.

So here we are, millions of us, standing on this beach in southern Florida, waiting for the asteroid to arrive.

SCENARIO

A "planet-destroying asteroid strike" is one of the most interesting and compelling doomsday scenarios that humans face, for these four reasons:

1 We know that it has happened before. The famous "asteroid that killed off the dinosaurs 66 million years ago" is the best-known asteroid strike; there have been many others.

2 We know that it will happen again. The only question is when.

3 Lots of serious scientists take this scenario seriously. Different countries and institutions have spent a large amount of time, effort, and money to understand the full extent of the threat.

4 An asteroid of sufficient size would kill half or more of all the life on our planet. Nothing would be the same afterward.

How do we know, with very high certainty, that a giant asteroid perhaps 10 to 50 miles (16 to 80 km) in diameter hit Planet Earth 66 million years ago? Several interlocking pieces of scientific evidence lead us to this conclusion. Worldwide fossil records show that a mass extinction event occurred at this time—in the fossil record, it's as if dinosaurs existed one day and were gone the next. Many other species, about 75 percent of life on Earth, were extinguished as well.

If we dig down through Earth's geological layers to this depth (or find the layers exposed in upheaved rock), there is a thin layer that contains an unusually high concentration of iridium, an element that is not common on Earth but often found in asteroids. This "iridium anomaly" was first identified and publicized in 1980 by a group that included Nobel prize–winning physicist Luis Alvarez and his son, geologist Walter Alvarez.

In addition to the iridium, geologists often find tiny glass beads called *tektites* in this layer. They are formed when an asteroid's impact melts Earth's crust and ejects the molten rock into the atmosphere or space. As the ejecta flies through the air, it cools into tiny spheres before falling back to Earth, like raindrops of glass. Tektite deposits were discovered to be particularly thick in Haiti, which indicated that the impact crater was likely nearby. In 1990, the crater was located off the coast of Mexico; it is now called the Chicxulub crater. Although we don't know the exact size of the asteroid that hit Earth 66 million years ago, it was a big one. The crater is 93 miles (150 km) in diameter.

Although this dinosaur-killing asteroid is the most famous, there have been many other asteroid strikes on Earth. The Vredefort crater in South Africa is 185 miles (300 km) in diameter and was created by an asteroid strike 2 billion years ago. The Chesapeake Bay crater is 35 million years old and 53 miles (85 km) in diameter. It is easy to see evidence of asteroid strikes on the moon because of all the craters that these strikes leave behind, especially on the far side of the moon. These craters on the moon last for eons because there is no rain or wind to erode them away.

What are the odds of another large asteroid strike on Planet Earth? Surprisingly high. The only question is when and where. We already know that there are tens of thousands of near-Earth objects (NEOs) that could hit our planet, and we discover 1,500 new NEOs each year right now. According to *Nature* magazine, the odds are 0.01 percent (a 1-in-10,000 chance) that an asteroid 300 feet (100 m) in diameter could hit Earth in any given year. An asteroid big enough to cause a mass extinction event, 1 kilometer (0.6 mile) in diameter or greater, has a one-in-a-million chance of hitting Planet Earth in any given year—a remarkably high probability given the consequences. Scientists calculate these odds by looking at several factors, such as the number of NEOs we have discovered and the historical rate of events occurring on the Earth, the moon, and other bodies in space.

A large asteroid strike is a worst-case-scenario for Planet Earth and all the life-forms it harbors. It will cause a combination of effects. There is the crater itself, many miles in diameter. Then there are the millions of tons of material ejected from the crater. If the asteroid hits the ocean, it can generate massive tsunamis (see page 150). The immense energy released by the impact will create a gigantic fireball, lighting forests and structures on fire. Because this would be such a massive blow, the impact will trigger seismic waves, earthquakes, and volcanic eruptions.

There are also wind and pressure effects. During any explosion, the blast pushes away a lot of air at very high speed, creating a blast wave (or shock wave). The blast wave consists of this mass of compressed air moving very fast. Part of the damage from blast waves comes from the wind-like effect. Another part comes

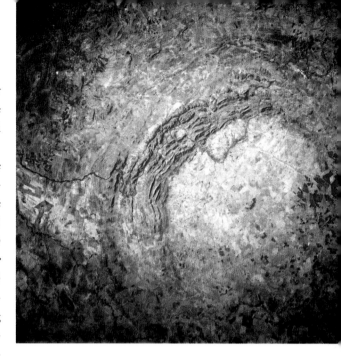

ABOVE Though the crater from the asteroid that struck present-day South Africa two billion years ago has been erased, the asteroid left behind a dome-shaped structure known as Vredefort Dome.

from overpressure, in which the air pressure is much higher than normal because of the compression effect. This blast effect was best demonstrated on a small scale in 2013, when an asteroid estimated at 66 feet (20 m) in diameter exploded over the Chelyabinsk Oblast region of Russia. It was a tiny event as asteroids go, but the blast wave from the explosion is what did the most damage. About 1,500 square miles (4,000 km²) were affected by this blast wave. It shattered windows and collapsed roofs in the affected area.

With an asteroid of sufficient size, the combination of all these effects is so catastrophic that most of the life on the planet is destroyed. Some of it may be destroyed immediately, or it may be destroyed by the severe, planet-wide winter weather that follows the impact. Sadly, even if you survive the immediate catastrophic effects of a large asteroid strike, they're not the worst part. The long-term effects for the rest of the planet are in fact more frightening. An unimaginable amount of soot, smoke, rock, water, dust, and sulfur will blow into the sky from the impact, contaminating the atmosphere and blotting out the sun for years. With no sunlight, plants will die, as will plankton in the oceans, and temperatures will drop significantly around the globe, leading to the collapse of the entire food chain.

As for people hunkering in bunkers, can they realistically ride out an asteroid strike underground? The answer will depend on whether they can endure the worst conditions on the planet's surface, which could remain dire for several years and possibly a decade or more. We don't really know how long it would take for the skies to clear and normal weather patterns and temperatures to reestablish themselves. We expect that surface temperatures would drop approximately 50°F (27°C) from the reduced sunlight, effects that could be long-lasting. This period of cold weather is known as "impact winter."

CATEGORIZING NEAR-EARTH OBJECTS

NASA divides NEOs that might endanger all or parts of the planet into three categories:

- **Objects between 140 meters (460 feet) and 300 meters (985 feet) in diameter:** Asteroids that might take out a state or a small country.
- **Objects between 300 meters (985 feet) and 1,000 meters (3,280 feet):** Asteroids that would have serious effects on whole regions, like a continent.
- **Objects 1,000 meters (3,280 feet) in size or larger:** Asteroids that would likely affect the entire planet, like the one described at the beginning of this chapter.

The interesting thing about NEOs is that we don't have anywhere near a complete catalog of them. Using telescopes and radar, NASA is discovering about 1,500 new objects per year. Perhaps a third of the total number has been discovered to date, meaning there could be as many as 40,000 still hidden and awaiting discovery.

BELOW Four of the largest detected asteroids, from left to right, are shown on a map of North America for scale: Ceres, Vesta, Pallas, and Hygeia.

Why are there near-Earth objects threatening Planet Earth? Mars and the moon don't seem like they will leave their long-standing orbits and aim themselves for a direct strike on Earth anytime soon. So, why do asteroids go rogue and create such a threat? Why don't asteroids stay in their orbits out in the asteroid belt (a region between the orbits of Mars and Jupiter) and leave us alone?

The first thing to understand is that there are untold millions of asteroids in the asteroid belt. There are more than a million with a diameter greater than 1 kilometer (0.6 mile). It is impossible to know the full scale of the asteroid population because smaller ones are difficult to detect from Earth.

There are many asteroids that exist outside the asteroid belt and many more that have orbits that bring them in and out of Earth's neighborhood in the inner solar system. Because of their proximity to Jupiter, asteroids can change orbit when Jupiter's gravity affects them. When we combine all these effects, we find tens of thousands of objects whose orbits will happen to cross Earth's orbit. If the path of an asteroid or comet happens to cross Earth's orbit at a time when Earth happens to be at the same spot, this is when a collision occurs.

Finding all these wayward asteroids that could potentially hit Earth is not an easy task. The asteroids are relatively small, they can be millions of miles away, and they are dark. The main technique currently used to detect them right now is Earth-based telescopes that look for changing amounts of reflected light as asteroids pass through space.

BELOW As part of the NEOWISE project, NASA launched the Wide-field infrared Survey Explorer (WISE), a space telescope, in 2009. WISE used infrared waves to detect and measure NEOs. This photo, captured by WISE, shows an NEO called 1998 KN3.

THE DESTRUCTIVE POWER
OF ASTEROIDS,
QUANTIFIED

It would seem obvious why an asteroid impact is such a devastating event, but just how much energy does a huge asteroid release? Let's consider our 25-mile-diameter asteroid as it hits the planet at 40,000 miles (65,000 km) per hour.

For the sake of this calculation, we'll assume that the asteroid is made of rock with the density of granite. Granite has a mass of 2.75 grams per cubic centimeter (0.10 pounds per cubic inch). A sphere that is 25 miles in diameter has a volume of 3.41×10^{19} cubic centimeters, so the asteroid would weigh roughly 10 trillion metric tons.

Now imagine those 10 trillion metric tons are traveling at 40,000 miles per hour (64,000 km/h, or 17,777 m/s).

The Kármán line, which is considered the boundary between the Earth and space, measures about 100 kilometers (67 miles) above the Earth. Traveling at 17 kilometers per second (11 miles per second), our asteroid's 10 trillion metric tons would pass through the entire atmosphere in only five or six seconds. There might be some hope that the asteroid would slow down as it entered the atmosphere, but at this scale, the atmosphere is essentially meaningless to the asteroid. When the asteroid strikes Earth, it will release about 16 million quadrillion (1.6×10^{24}), or 16 sextillion, joules of energy.

Let's put this amount of energy into perspective:

- A kilogram (2.2 pounds) of TNT contains 4.2 million joules (4.2×10^6 joules). This is roughly equivalent to the energy from lighting a soda can filled with gasoline on fire.

- A metric ton of TNT contains 4.2 billion joules (4.2×10^9 joules). This is equivalent to lighting 100 gallons of gasoline (380 l) on fire. Imagine the energy in a typical car's gas tank when it explodes, then multiply that by five.

- The atomic bomb dropped on Hiroshima, Japan, in World War II released 63 trillion joules (6.3×10^{13} joules), the equivalent to 15,000 tons of TNT. The blast and fire from this bomb leveled everything in a 2-mile-diameter (3.2 km) circle.

- Russia's Tsar Bomba, the largest nuclear bomb ever detonated, contained 210 quadrillion joules (2.1×10^{17} joules), the equivalent energy of 50 million tons of TNT.

When our hypothetical asteroid strikes, it will be roughly 10 million times stronger than the largest nuclear bomb ever detonated, equivalent to something like 500 trillion tons of TNT.

RIGHT This sequence of images shows the effect of a nuclear blast on a wooden house. In photo A, the house receives illumination from the blast. The temperature of the explosion is millions of degrees. Photo B shows the effects of the resulting thermal radiation hitting the house. In photos C through F the house is blown apart by the blast wave. All of this happens within a few seconds, and the severity of the effects depends on the building's distance from the blast site.

CLOSE CALLS WITH ASTEROIDS

- On November 8, 2011, a large (1,300 feet, or 400 m, in diameter) asteroid named "2005 YU" came within 200,000 miles (321,869 km) of Planet Earth. It had been discovered in 2005. Had this asteroid struck Earth, it would have had enough impact energy to take out a whole continent.

- On January 2, 2018, an asteroid named "2018 AH," very similar in size to "2019 OK" (next on this list) came within 184,000 miles (29,6119 km) of Earth.

- On July 25, 2019, a 300-foot (90 m) diameter asteroid named "2019 OK" came within 45,000 miles (72,420 km)

of Earth. Keep in mind that Earth is 8,000 miles (12,875 km) in diameter, so this is an extremely close approach in astronomical terms. This asteroid had not been detected ahead of time, so no one knew it was coming. Had it hit Earth, it would have struck with the energy of a nuclear bomb. If it had struck a city, the city would lie in ruins.

- In 2028, an asteroid known as "2001 WN," half a mile (0.8 km) in diameter, will make a close approach to Earth. If it were to hit Earth, it could conceivably destroy a region the size of Europe.

PREVENTION

To prevent an asteroid strike, humans need years of advance warning. Getting to this point would mean that we need to discover and catalog all the NEOs that endanger the planet. (Keep in mind that humanity has yet to catalog even half of the NEOs that could possibly threaten us.) We then need to predict their trajectories and highlight any that will come close to the Earth.

This scenario is so compelling that NASA created the Planetary Defense Coordination Office (PDCO) in January 2016 to get a handle on the problem. The PDCO has three goals: finding and tracking near-Earth objects (NEOs) that might hit the planet; sorting out which objects are on a collision course with Earth by calculating their

orbits and trajectories; and figuring out the best way to respond to an impact threat. The International Asteroid Warning Network (IAWN) established by the United Nations does similar work on a global scale. The European Space Agency has the Space Situational Awareness (SSA) Programme.

The most difficult, and crucial, task is to eliminate the possibility of impact from threatening NEOs. There are a number of theoretical options, most of which involve changing the asteroid's trajectory enough to miss us. If caught early enough, an asteroid that is on a collision course with Earth can turn into a miss with just a tiny change in its trajectory, because the asteroid is flying through such large distances before it arrives. Think about

THE DOOMSDAY BOOK

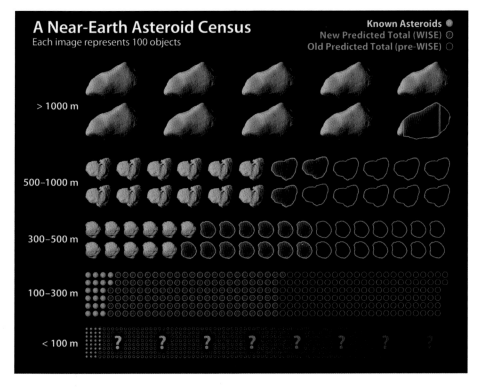

A Near-Earth Asteroid Census
Each image represents 100 objects

Known Asteroids ●
New Predicted Total (WISE) ○
Old Predicted Total (pre-WISE) ○

> 1000 m

500–1000 m

300–500 m

100–300 m

< 100 m

ABOVE Using data collected by WISE, NASA conducted an asteroid census and estimated the number of different size asteroids in our solar system.

your car traveling on a straight road. If the steering is off by even the tiniest amount, then the car will eventually drive off the road if you travel far enough.

One way to accomplish this is to hit the asteroid with a nuclear weapon well before impact. The colossal force of the blast along with the material ejected would change the asteroid's course very slightly. Hitting the asteroid with a kinetic impactor could be another plausible option. The kinetic impactor would be a heavy satellite that is accelerated to a high speed and then slammed into the asteroid. Given the utterly enormous mass of a planet-threatening asteroid, the impact would only have a tiny effect on the trajectory. But if done years in advance of impact, it might be enough to direct the asteroid ever so slightly elsewhere. NASA is preparing to try out this strategy with an experiment called DART, scheduled for 2022. Alternatively, some kind

of rocket engine powered by either solar panels or a nuclear power plant could be attached to the asteroid. The engine would then push or pull the asteroid to a new trajectory that misses Earth.

Rather than changing the asteroid's trajectory, we could pulverize the asteroid or dismantle it and carry it away. There are companies today that are looking into mining asteroids for their valuable metals. The idea here is to simply mine an incoming asteroid until it disappears or is greatly diminished.

To make these approaches more realistic, three things would have to happen: First, we need a complete catalog of all NEOs. Second, humans would need years of warning. Third, all four of the techniques mentioned just above would have to be ready and waiting to deploy so there's no delay once a dangerous asteroid is detected. The truth is, we might need to try all these approaches to get the job done.

SUPERVOLCANO ERUPTION

THREAT LEVEL: WORLD

If you were born in the 1960s or earlier, chances are that you vividly remember the eruption of Mount Saint Helens on May 18, 1980, in Washington state. The volcano had "woken up" about two months earlier, venting steam and causing thousands of small earthquakes. There were also cracks and fissures on the north face of the mountain that indicated building pressure; that side of the mountain started bulging outward.

When the whole thing blew on May 18, it was spectacular—more of an enormous bomb blast than an eruption. The first sign of the eruption was an earthquake that began at 8:32 a.m. Seconds later, the largest landslide ever recorded occurred as the bulging north face of the volcano collapsed. Nearly a cubic mile of rock and dirt slid down and, with all that weight out of the way, the volcano exploded.

The energy of the blast was estimated to be equivalent to a 24-megaton nuclear bomb. It blew ash and rock 80,000 feet (24 km) into the air, and this debris rained down on eleven states and parts of Canada. Dust from the eruption eventually circled the globe. Even today, you can see the collapsed face on the north side of the volcano and the debris field spread out below it. Mt. St. Helens is in a remote area, but even so 57 people died. Two hundred thirty square miles (370 km²) of forest were mowed down or air-fried to a crisp.

Mt. St. Helens may have been an enormous volcanic eruption and may have caused the largest landslide in recorded history, but it pales in comparison to the 1883 eruption of Krakatoa in Indonesia. Krakatoa created the loudest sound in recorded history. People heard it from as far as 3,000 miles (about 4,828 km) away. Imagine a sound in New York City that people hear in Los Angeles. That is incredible, but it happened with Krakatoa. All over the world, barometers detected the pressure wave from the blast. In fact, the pressure wave from this gigantic explosion circled the globe multiple times.

The eruption of Krakatoa was approximately eight times more powerful than Mt. St. Helens. Whereas the Mt. St. Helens eruption blew one cubic mile (1.6 km³) of material skyward, Krakatoa's eruption launched six cubic miles (9.7 km³) of rock and debris into the air. Thirty-six thousand people in the area died either directly from the blast and pyroclastic flows or in the ensuing tsunamis (see page 150).

The 1883 eruption of Krakatoa left a mark in the skies for years. The dust in the upper atmosphere produced vivid red sunsets for months and helped scientists identify and discover jet streams. It is thought that these Krakatoa-induced sunsets may have inspired the red sky in *The Scream* (1893) by Edvard Munch.

RIGHT The eruption of Mount St. Helens blasted ash 80,000 feet (24,000 m) into the air.

An enormous amount of sulfur dioxide from the eruption entered the stratosphere and lingered there for years. Sulfur dioxide increases the amount of sunlight reflected back into space and therefore led to global cooling that lasted at least five years. The summer following the eruption was cooler than usual, and rainfall in places as far away as southern California was significantly greater than normal.

The eruptions of Mt. St. Helens and Krakatoa certainly were catastrophes, but they are not on the doomsday scale. For a doomsday scenario—an eruption that would wreak the most havoc imaginable on Planet Earth—we need to talk about supervolcanoes. When a supervolcano erupts, it is capable of moving more than 1,000 cubic kilometers (621 cubic miles) of debris. There are a number of known potential supervolcanoes on Earth today, each one of them with the potential to disrupt nations and even the planet as a whole.

ABOVE A pyroclastic flow is an unbelievably powerful and unexpected side effect of many volcanic eruptions. It is a combination of superhot gases, ash, crumbled rock, and so on, that flows downhill much too fast to outrun, like a super-heated avalanche of debris. INSET An illustration showing the eruption of Krakatoa in 1883.

Scientists have identified twenty different super-volcanoes around the world. They all are unique in their age, geography, and frequency of eruption. One of the best known supervolcanoes lives under Yellowstone National Park in Wyoming, so let's use it as a typical example of the supervolcano genre.

The Yellowstone supervolcano has erupted multiple times in the past. Examples of prior eruptions include:

- The Huckleberry Ridge eruption: occurred 2.1 million years ago, ejected 2,200 cubic kilometers (1,367 cubic miles) of material into the atmosphere.

- Mesa Falls eruption: occurred 1.3 million years ago, ejected 280 cubic kilometers (174 cubic miles) of material into the atmosphere.

- Lava Creek eruption: occurred 640,000 years ago, ejected 1,000 cubic kilometers (621 cubic miles) of material into the atmosphere.

ABOVE The Grand Prismatic Spring is the largest hot spring in the United States. Yellowstone's most vivid landmarks, such as hot springs, bubbling mud, and geysers like Old Faithful, are powered by the thermal energy that comes from the Yellowstone Caldera.

THE WORLD'S TOP 10 SUPERVOLCANOES

VOLCANO	LOCATION	LAST ERUPTED
Aira Caldera	Japan	22,000 years ago
Campi Flegrei	Italy	480 years ago
La Garita Caldera	United States	28,000,000 years ago
Long Valley Caldera	United States	767,000 years ago
Mount Tambora	Indonesia	200 years ago
Paekdu Mountain	North Korea	1,060 years ago
Taupo Caldera	New Zealand	26,500 years ago
Toba Caldera	Indonesia	74,000 years ago
Valles Caldera	United States	68,000 years ago
Yellowstone Caldera	United States	640,000 years ago

If these three eruptions represent a temporal pattern, then the next eruption could happen any day now (in geological terms). Like the eruption of Mt. St. Helens, there would likely be some advance notice that something is happening in the Yellowstone caldera. There would be earthquakes, possibly some bulging evident in the terrain, and increased magma activity under the surface. Then, one day, the underground pressure would build to the point where the volcano explodes.

Any supervolcano eruption has the potential to be massive, easily blowing thousands of cubic kilometers of material skyward. The explosion then has major side effects:

- **What goes up must come down.** The millions of tons of ash, rock, and dust ejected into the sky will rain down onto homes, fields, and rivers.

- **Massive amounts of sulfur dioxide are blown into the stratosphere.** This will cool the planet for years to come.

- **The explosion of the supervolcano itself will be devastating.** Many people will die immediately from the blast and the pyroclastic flows that follow. Cities near the eruption are likely to be entombed in much the same way Pompeii was when Mt. Vesuvius erupted in 79 CE.

But then, after the immediate effects, there will be long-term consequences as well. The amounts of material and sulfur expelled by a supervolcano will be at least 100 times greater than what Krakatoa released. This ash creates many problems. It contains fine glasslike particles that damage the lungs, making it dangerous for humans and animals to breathe it in. Humans will need to wear masks to filter out the dust when they are outside, while most farm animals and wildlife will die. These airborne particles will also disrupt air travel, possibly for months. Volcanic ash destroys jet engines through both abrasion and clogging.

As the ash and debris start falling back to Earth, more problems arise. The weight of the ash will cause roofs to collapse, and as much as a foot of ash will accumulate on roads in a place like Denver, which is about 400 miles (644 km) away from the volcano. The humans who survive the building collapses and the air pollution will be trapped. There is no way to drive in this much ash. When ashfall hits major agricultural regions, such as America's "breadbasket"—the areas in the Midwest that grow the wheat and corn that feeds hundreds of millions of people—farming may be impossible for years, depending on ash thickness. The ash would cover everything, crushing existing crops, burying pastures, and making it difficult or impossible to plant seeds. Rivers

WHAT IS THE DIFFERENCE BETWEEN A NORMAL VOLCANO AND A SUPERVOLCANO?

How do we know that the Yellowstone caldera in the United States or Mount Tambora in Indonesia are supervolcanoes? The answer is straightforward: to be classified as a supervolcano today, there needs to be evidence of an eruption in the past that yielded 1,000 cubic kilometers (621 cubic miles) or more of ejecta. In most cases, when a volcano releases this much material, it leaves behind a noticeable crater, called a caldera. To get some perspective on the size of a supereruption, imagine taking the entire state of Rhode Island, which is 3,340 square kilometers (2,075 square miles) in size, and digging a hole encompassing the entire state that is 300 meters (985 feet) deep. This utterly enormous crater is what 1,000 cubic kilometers would look like.

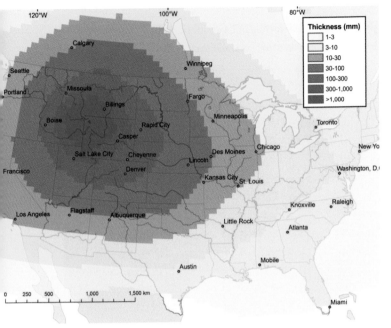

Thickness (mm)
- 1–3
- 3–10
- 10–30
- 30–100
- 100–300
- 300–1,000
- >1,000

ABOVE This map shows the areas where ash would fall if the supervolcano at Yellowstone National Park were to erupt. The blue, purple, and pink areas would see the highest amounts of ashfall.

will clog when stormwater runoff and rain carry ash from nearby hillsides into the water. Food and water will likely be scarce as power plants shut down.

Then the worst part unfolds: it is quite likely that volcanic winter envelopes the planet for years due to sulfur dioxide injected into the stratosphere and to the immense amount of dust lingering in the air. Temperatures will drop, leading to widespread crop failures. This problem arises because sulfur dioxide reflects sunlight back into space. A great example of this phenomenon happened when Mount Tambora in Indonesia erupted in 1816, known as "The Year without a Summer" in the northern hemisphere. In New England, there was snow and frost in June. Europe—especially Britain, Germany, and Switzerland—faced famine conditions. By supervolcano standards, this was a tiny eruption of "only" 40 cubic kilometers (25 cubic miles). A supervolcano would eject at least fifty

times more material and sulfur dioxide into the atmosphere.

We have two events that tell us a little about what a full-scale supervolcanic winter may entail. About 75,000 years ago, the Toba supervolcano erupted in present-day Sumatra, Indonesia. By studying human DNA, scientists theorize that a "bottleneck" in human population occurred at about that time. The world's population of humans may have shrunk to as few as 3,000 people, suggesting that the effects of this eruption may have been bad enough to nearly extinguish the human species. In 1257, the Samalas volcano erupted near Indonesia. Perhaps twice as strong as the Krakatoa eruption, it lowered temperatures in Europe enough to cause a famine and may have triggered a mini-ice age that lasted for decades.

A volcanic winter could last for five years or more. A mass extinction event similar to the one described in the chapter on asteroids (see page 122) is quite likely if a supervolcano eruption is large enough. Half or more of the life on Planet Earth could be lost.

ABOVE Ashfall from the 2006 eruption of Mount Merapi in Indonesia destroyed this house in central Java, along with entire villages

ABOVE Lava flows from Kīlauea Volcano in Hawaii. This lava eventually reaches the ocean and creates hundreds of acres of new land in the process.

To understand volcanoes, we need to understand the layers that make up Earth. The inner layers of our planet, the upper mantle and lower mantle, are extremely hot—hot enough to melt rock. So these layers are primarily molten rock, which we call magma. When we see red-hot lava flowing above ground, it is magma that has escaped the upper and lower mantles.

The outer crust consists of the tectonic plates that can move, albeit slowly, on the more fluid magma layer below it. The rigid, cooler tectonic plates are called the lithosphere, and they move over the more fluid and warmer asthenosphere. The asthenosphere is defined by its temperature, being made up of rock that is 1,300°C (2,372°F) and higher.

There are about 1,500 active volcanoes on the planet. The majority of them form at sites where tectonic plates come together, with one plate pushing under the other. These sites are called subduction zones.

Volcanoes can also form over "hot spots," random places where magma has pushed through the middle of a tectonic plate. The reasons for why this occurs are not well understood yet. The Yellowstone supervolcano and the volcanoes that formed (and are still forming) the Hawaiian islands are both over such hot spots.

In the case of Yellowstone, the hot spot has created a magma chamber that is very close to the Earth's surface. Seismic tomography has allowed us to predict what this chamber looks like. So far, we know there

are two magma areas underneath Yellowstone—the magma chamber, and a much larger magma reservoir beneath it. We know that they are stable; magma is not pushing into them or leaving—for now.

When the magma beneath a volcano starts pushing toward the surface, this could mean trouble. This type of change is what caused the eruption at Mt. St. Helens, and it is why people get nervous when there are earthquake swarms around Yellowstone. The Yellowstone Volcano Observatory keeps track of any earthquakes, as well as the elevation of the area, in order to detect bulging. The observatory issues monthly reports based on its findings. For example, one earthquake swarm began in December 2008. Over a period of two months, it generated about 500 small earthquakes before subsiding. There is a possibility that one day, a swarm will start without subsiding: instead, it will accelerate and turn into an Earth-shattering eruption.

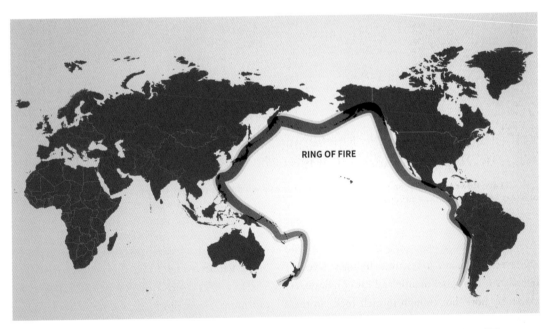

ABOVE The various subduction zones around the Pacific tectonic plate form an area called the Ring of Fire, which contains more than 400 volcanoes.

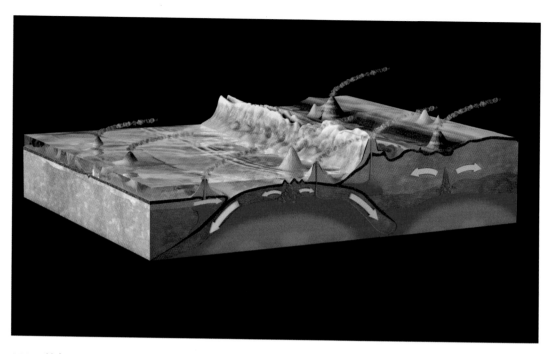

ABOVE Volcanoes can form at subduction zones, where one tectonic plate submerges under another, or at hotspots that form in the middle of tectonic plates.

SEISMIC TOMOGRAPHY

A technique called seismic tomography has allowed us to visualize what lies beneath volcanoes quite accurately. Seismic tomography relies on seismic sensors that detect earthquakes very precisely. There are now thousands of these sensors (often called seismometers) located all over the planet. By using this multitude of sensors, scientists can precisely locate the epicenter of every earthquake, and then calculate the exact travel time of the earthquake from the epicenter to each seismic sensor. By combining thousands of measurements and looking at the different travel times, it is possible to determine what must be underground, whether it's solid rock, molten rock, layers between different kinds of rock, or something else. Since every eruption creates an earthquake-scale event, and is often preceded by smaller earthquakes, earthquake science and volcano science are tightly linked.

THE DOOMSDAY BOOK

Humans don't have a lot of control over tectonic plate movements, subduction zones, hot spots, or magma flows. While it is possible to imagine human beings coming to the point where we could stop an asteroid from hitting the planet (see page 130), the activities inside the planet are not as observable or controllable.

Nonetheless, NASA has a proposal that might help prevent the Yellowstone supervolcano and possibly others from erupting. The idea is to cool off the magma chambers and essentially "freeze" them in place by injecting so much cool water that the heat of the magma is siphoned away.

If this extracted heat can be tapped as a geothermal energy source, there is potentially enough energy available there to power the United States. The setup of this geothermal power plant would not be tremendously complicated. After water is injected into the ground, the magma turns it to steam, which in turn drives a turbine and a generator that creates electricity. In the process, the magma cools and eventually solidifies, becoming much less of a threat.

A geothermal plant like the one proposed by NASA would face challenges. Many people are not keen on dropping geothermal power plants into national parks, such as Yellowstone, for aesthetic reasons. Once the magma cools down, all of its hydrothermal features would likely stop working. Hawaii faces a similar situation, where religious and aesthetic concerns have blocked the use of geothermal power. On the other hand, Iceland has had good luck with geothermal power. Nearly all the homes in Iceland have geothermal heating, and much of Iceland's electricity is geothermally powered as well.

If we cannot avert a supervolcano disaster, what can we do to prepare? To help people survive the long-term effects of an eruption, we could make sure that people have plenty of breathing masks along with goggles on hand ahead of time, to avoid breathing in the harmful ash. Stockpiling food and water on an individual, municipal, or national level could also prevent deaths related to a volcanic winter. We have seen enough scenarios like a volcanic winter in the past, where climate conditions have deteriorated to the point where food cannot be grown for several years. Setting aside a three-year or even five-year supply of food for a whole nation would be complicated and expensive and would require a lot of cooperation, but we would be better prepared for a variety of doomsday scenarios, from supervolcano eruptions to asteroid impacts, if we did.

ABOVE Bathers enjoy the warm waters of the Blue Lagoon, a popular geothermal bath resort in Reykjavík, Iceland. The volcanic activity of this region also provides the primary energy source for the Svartsengi geothermal power plant, visible in the background.

EARTHQUAKES

THREAT LEVEL: CITY

Imagine that you are one of the more than 7 million people living in the San Francisco metropolitan area (SFMA). You certainly know that an earthquake is possible. Just about everyone is aware that the SFMA sits very close to the famous San Andreas Fault. And there are several other fault lines near the region, such as the Hayward and San Gregorio faults. A fault line is a place where two tectonic plates meet and is often a spawning ground for earthquakes.

What's more, San Francisco has a history of destructive earthquakes measuring a magnitude of 7 or greater on the Richter scale, including the San Andreas Earthquake (magnitude 7.0) in 1838 and the Hayward Earthquake (magnitude 7.0) in 1868. The twentieth century saw two major events: the Great San Francisco Earthquake (magnitude 7.8) in 1906 and 1989's Loma Prieta earthquake (magnitude 7.1). There are smaller events all the time.

San Francisco's 1906 earthquake basically leveled the city, destroying about 80 percent of its buildings either in the earthquake itself or in the fires that followed. The fires started primarily where gas lines ruptured and quickly spread from there because most of the buildings were wooden. In 1906, about 400,000 people lived in San Francisco—3,000 of them died, and nearly everyone lost their homes. This was doomsday at the city level.

There is a famous fence on the "Earthquake Trail" outside San Francisco that once sat right on the San Andreas Fault. When the fault line gave way in the 1906 quake, the North American tectonic plate moved 16 feet (4.5 m) relative to the Pacific tectonic plate in about a minute. That straight fence became two fences 16 feet apart.

It is almost impossible to imagine so many square miles of land—terra firma!—moving 16 feet in a minute, but it was this shift that caused the earthquake, knocked down buildings, fractured gas lines, and led to so much loss of life.

The 1989 earthquake that struck San Francisco was not quite as strong as the one in 1906, but San Francisco was also far more developed. All kinds of things, from buildings to bridges totaling $5 billion, were damaged, even with earthquake-proofing measures. But what no one anticipated was the destruction that happened down by the bay. Large areas of new land had been created by filling in an old lagoon in the Marina district. When the ground shook, some of this new land liquefied, causing buildings to collapse.

RIGHT More than 28,000 buildings were destroyed in the 1906 Great San Francisco Earthquake and the subsequent fires.

Buildings falling, bridges collapsing, pipelines rupturing, soil liquefying, people dying—these are all features of any powerful earthquake. If the quake is strong enough—if one tectonic plate moves far enough relative to another when a fault line lets go—there is nothing humans can do to make a city completely safe.

SCENARIO

On the doomsday scale, earthquakes generally rank as a less severe disaster, because they are localized. Even so, a big earthquake can impact, and possibly ruin, a large city. In the case of the earthquake that struck Haiti in 2010, over 100,000 people died in and around the city of Port-au-Prince. Hundreds of thousands of homes and buildings collapsed, primarily because of substandard building codes and masonry construction that never accounted for possible earthquake damage. It was an apocalyptic situation for Haiti, one from which the country has yet to completely recover.

The Kobe Earthquake in Japan in 1995 is another example. This earthquake happened near the highly developed city of Kobe (population 1.5 million) and was surprisingly destructive. Damage estimates hovered around $200 billion and included major repairs to highways, rail lines, and approximately 400,000 buildings.

So, why did a developed country like Japan, which has a high frequency of earthquakes, end up with so many buildings and structures damaged in 1995? The reason: It wasn't until the 1970s and 1980s that engineers and architects truly started to understand how to build earthquake-proof buildings. Structures in Kobe built *after* Japan's building codes were revised in 1981 did well; structures built prior to 1981 proved far more vulnerable. This difference was especially evident with the Hanshin Expressway (Route 43), which opened in 1962. There simply was not enough steel and strength in the support columns to resist the motion caused by the earthquake, and the highway collapsed in spectacular and catastrophic fashion (see image on page 121).

WHAT IS THE RICHTER SCALE?

We need some way to talk quantitatively about the strength of an earthquake. Back in 1935, a scientist named Charles Richter recognized this need and wrote a paper about this now-famous scale. On his scale, each number represents a tenfold increase in the earthquake's intensity. For example, a magnitude 7 earthquake is ten times stronger than a magnitude 6 earthquake. A magnitude 3 earthquake is very minor—most people can feel it, and objects that are not bolted down may shake slightly. When an earthquake reaches a magnitude of 7 and above, buildings start falling down.

ABOVE The magnitude 7.0 earthquake in Haiti in 2010 caused billions of dollars in damage and killed over 100,000 people.

Another big problem in the Kobe earthquake was soil liquefaction, where unstable soils start acting like liquids when shaken. Liquefaction can be an acute problem for buildings and roads constructed on fill-in tracts of land. Near Kobe, construction of the artificial island known as Port Island started in 1966 and finished, coincidentally, in 1981. To build this island, the city dismantled Mount Yoko near Kobe and dumped the dirt and rock from the mountain into the bay to form the land. It was a huge project, one that created approximately 1 square mile (1.6 km²) of new land. When the earthquake came, the island suffered a number of failures as the unstable soil liquefied. Large cracks opened up in the ground. Cranes toppled as the ground beneath them weakened.

Earthquakes can also portend even greater destruction. Earthquakes frequently precede a volcanic eruption (see Supervolcano Eruption, page 132). Earthquake swarms often indicate the movement of magma into a volcano's magma chamber. An earthquake, particularly those underwater, can also cause tsunamis. This is exactly what happened in the 2004 Indian Ocean tsunami (see page 150), an incredibly deadly event that killed hundreds of thousands of people across many countries.

Earthquakes sometimes accompany fracking, or more specifically the wastewater disposal associated with oil and gas wells. It's common practice in the industry to dispose of contaminated wastewater by injecting it into the ground under high pressure. The earthquakes that occur as a result of the added water are called "induced" earthquakes. In Oklahoma, earthquakes have multiplied by a factor of 40 over the last decade as fracking and oil production have increased.

If the epicenter of an earthquake occurs in the middle of nowhere, it may affect nobody. Earthquakes often affect just one city near the epicenter of the quake. But there can be cases where earthquakes have larger-scale effects. When an earthquake lifts a tectonic plate in the middle of the ocean, the resulting tsunamis can affect many cities. And some fault lines are so long and so deep that they can store enough energy to create magnitude 9 or higher earthquakes. One such fault line exists in Japan, and we saw the effects in 2011 when the Great Tōhoku Earthquake hit 9.0 on the Richter scale. The resulting tsunamis devastated several Japanese cities. The Cascadia fault line off the coast of the Pacific Northwest region of the United States could potentially store up enough energy to release a magnitude 9.3 earthquake. Because it is 60 miles (97 km) off the coast, it could create significant tsunamis as far away as Japan, as well as impacting large US cities like Seattle and Portland.

SCIENCE

Earthquakes occur due to the Earth's moving tectonic plates. These plates can collide with one another in a process that creates mountain ranges. They also slide under other plates (the places where this occurs are called subduction zones) or slide past one another at their junctions, as with the San Andreas Fault. The fastest-moving tectonic plates can move up to 150 millimeters (6 inches) per year, but average rates of movement tend to be slower. The rate seen in the San Andreas Fault is about 35 millimeters (1.4 inches) per year.

What causes enormous, extremely heavy pieces of land—some the size of continents—to move? The answer: magma under the plates moves due to convection, or heat transfer that results from movement of fluids, and the magma's motion drags the plates along for the ride. Hotter magma wants to rise to the surface, while cooler magma wants to sink toward the Earth's core, and the circulating effect created by these movements animates the plates. The magma follows the same sort of convective motion you see inside a lava lamp, where the blobs at the bottom, near the warm light, heat up, expand, and rise. Meanwhile, the blobs at the top cool down, contract, and sink back down.

Earthquakes occur in response to the plates' motion relative to one another at the places where they meet. Take the San Andreas Fault. Along its fault line, the Pacific tectonic plate and the North American tectonic plate are moving in different directions. Relatively speaking, the Pacific plate is moving north, and the North American plate is moving south.

Let's imagine that we could inject a perfect lubricating layer into the San Andreas Fault where the two plates meet. The two plates would slide evenly and freely past each other at a rate of about 35

Intrusion of Magma pushes the Plates away

Plates flow on the Convection currents

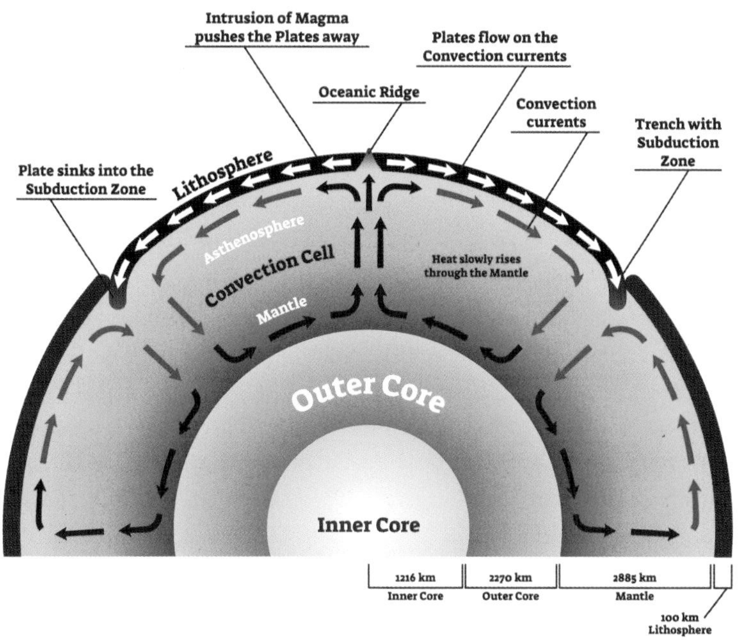

Oceanic Ridge

Convection currents

Trench with Subduction Zone

Plate sinks into the Subduction Zone

Lithosphere

Asthenosphere

Convection Cell

Heat slowly rises through the Mantle

Mantle

Outer Core

Inner Core

1216 km Inner Core	2270 km Outer Core	2885 km Mantle

100 km Lithosphere

ABOVE Convection currents cause Earth's tectonic plates to shift. There are often earthquakes where adjacent plates submerge or grind against each other.

millimeters (1.4 inches) per year. Every ten days, we would notice movement of about 1 millimeter (about $\frac{1}{32}$ inch) between the two lubricated plates, eliminating the risk of earthquakes along the San Andreas Fault.

Unfortunately, the two plates do not move freely past each other. Instead, they grind against each other along hundreds of miles of rock. The rocky edges of the two plates can lock together, sometimes for years, and pressure builds. When there is enough pressure, the two plates will break free at a weaker point in the fault line and slide all at once. If the plates of the San Andreas Fault have locked, with pressure building for ten years, and the plates break free at a weak point, the plates will move 350 millimeters (or about 1 foot) in just a few seconds. Now imagine the entire city of San Francisco moving

a foot south in, say, 20 seconds. It sounds unreal, but this is exactly what happens in an earthquake. If the plates lock together for a century, the whole city could move 10 feet (3 m). This sudden shaking would destroy buildings, bridges, and towers if they are not designed correctly.

One earthquake can sometimes beget another. Terms like fore-shock and after-shock can be used to describe this phenomenon. There are also swarms of earthquakes that can occur. All these variations come from the way a fault line ruptures. It might rupture all at once, leading to one big earthquake. It might have a small rupture that gets things started, followed by a big rupture, or vice versa. Or the fault might break slowly, in a series of little steps over the course of a few days, leading to a swarm.

Are human beings ever going to stop earthquakes? Probably not. If we could completely and perfectly lubricate all of the planet's fault lines, then perhaps, but this is utterly impossible today.

What humans can do is to better design our cities and structures to survive earthquakes unscathed. At this point, engineers have a good handle on how to do this (see page 149), and new structures can be designed and built to be earthquake-proof. The remaining problem is older structures that exist in earthquake-prone areas, as was highlighted so vividly in the Kobe earthquake.

What do we do about older buildings? One approach is to simply replace older structures where possible. Another idea is to reinforce them. For example, some older concrete bridge pillars in Seattle have been wrapped in steel to make them more rigid and earthquake-proof. A third approach is to lift buildings up and put them on isolation bearings when possible—a method called base isolation. Though this may sound extreme, it has been done a number of times when it made sense financially. For example, engineers added isolators to the city hall in Los Angeles to make it more earthquake-resilient.

When proper earthquake preparations are successfully and comprehensively implemented, an entire city could potentially experience an earthquake and come through it largely unscathed. If done on a global scale, earthquake-proofing measures could mark the end of this particular doomsday scenario except in the most extreme cases, or in cases that induce tsunamis (see page 150).

BELOW The sections of the Trans-Alaska Pipeline located near the Denali fault line are built in a zigzag to help it survive shifting land along the fault.

DESIGNING EARTHQUAKE-PROOF INFRASTRUCTURE

There are a few general approaches to designing buildings that can withstand earthquakes. With the first approach, engineers make the building's structure rigid and strong enough to withstand the shaking and stresses caused by the ground moving. This added strength prevents the structure from deforming or collapsing. Wooden buildings are reinforced with steel straps and brackets. Steel skyscrapers are built with extra bracing and steel, and concrete bridge pillars are built with extra steel in the interior or on the exterior. These kinds of modifications can be tested either with models or full-size buildings mounted on enormous shaking tables or with computer simulations and a mathematical technique called finite element analysis.

For the second approach, engineers disconnect the structure from the ground to some extent, so that when the ground shakes, the structure doesn't need to move quite as much. There are a wide variety of ways to allow structures to either fully or partially disconnect when the ground moves. One of the more extreme examples can be found on the 3-kilometer-long (1.9-mile-long) Rio–Antirrio bridge, which is supported by four enormous pillars. Located in Greece, the bridge has pillars that are not anchored into bedrock as you'd normally expect. Instead, they simply rest on a layer of gravel on the seafloor, with 90-meter-wide (295-foot-wide) bases to support the load. They are not formally connected to the ground in any way. During an earthquake, the ground can slide around under the pillars, while the pillars mostly stay where they are.

Other earthquake-proof structures around the world rely on spherical sliding bearings. As the ground moves, the bearing material slides easily over the surface of the spherical component. When the earthquake stops, the bearing re-centers itself as it slides back down into the depression. An elastomeric bearing, which consists of layers of rubber that can easily flex in response to an earthquake, provides similar protection from earthquake-induced damage.

Engineers face additional challenges when linear structures like roads, railroad tracks, and pipelines have to cross fault lines that are slipping. Sometimes there is nothing to be done except to repair things after an earthquake, as is often the case with railroad tracks.

In other cases, engineers can anticipate slippage. The best example of this can be seen in the section of the Alaska Oil Pipeline that crosses the Denali Fault line. The pipeline was built in a zigzag shape to give it extra length and flexible joints. In addition, Teflon® sliders disconnect the pipeline from the ground. The Teflon sliders are placed over slippery rails so when the ground under the pipeline shifts, the pipeline can accommodate the reconfigured landscape.

BELOW The pillars of the Rio–Antirrio bridge are not anchored to the ground but instead are constructed so the ground can slide underneath them, allowing the bridge to withstand an earthquake of 7.5 on the Richter scale.

SUPERTSUNAMIS

THREAT LEVEL: CONTINENT

When we felt the earthquake, it was unbelievable and seemingly impossible. How can the whole world move like this? Plus, it lasted several minutes, shaking the buildings, knocking things over in the bar, causing our poolside chairs to rattle. I have never experienced an earthquake, or anything like this, and it makes my stomach lurch. There is something completely unnatural about the Earth itself shuddering and convulsing. We expect our planet to be a rock-solid place, and an earthquake makes a mockery of this expectation, especially when the shaking lasts so long.

But then it is over—and it is as though nothing has happened. It is the day after Christmas in the year 2004. Sitting poolside at a beautiful resort near Banda Aceh, a city in northern Sumatra, we continue to enjoy our drinks. The sun is still out. The children return to their play. All the structures around us are still standing. The birds continue singing. I have heard about earthquakes destroying whole cities, but this one does not seem harmful at all. Life goes on.

Unfortunately, we have no idea that a chain of events has been irrevocably set in motion by this earthquake, and we are about to experience those events in the worst possible way. This is no normal earthquake. It will later be classified as an "undersea megathrust earthquake" with a 9+ magnitude on the Richter scale. Occurring about 100,000 feet (30,480 m) under the ocean just north of Banda Aceh, this earthquake is so powerful that hundreds of miles away, buildings are shaking in Bangkok, Thailand. In addition, this earthquake has caused a gigantic tectonic plate deep under the ocean nearby

to quickly shift upward by as much as 6 feet (2 m) in a matter of seconds. This upward movement of the tectonic plate is where the term *megathrust* comes from, and this upward movement is also about to kill a great many people.

This megathrust will not only force the plate upward: it also lifts the many cubic miles of seawater that are right above the plate. With this many square miles of the ocean floor having just relocated themselves, all of this newly uplifted water has to go somewhere, so it starts spreading out in all directions from the earthquake's epicenter and in particular toward Banda Aceh, the closest major city. Over time, the uplifted water will crash upon many shores in the form of tsunamis.

Twenty minutes after the earthquake, Banda Aceh is the first to see a slice of this displaced water. Its people can look out at the ocean and, in the

RIGHT In Banda Aceh, Indonesia, the Indian Ocean tsunami that struck in 2004 destroyed thousands of houses.

distance, the incoming wave is easy to see. Several waves, one as high as 30 feet (9 m), will rush onto the shore, flooding inland up to 2 miles (3.2 km).

The force of the water will pick up every loose thing in its path, including cars, appliances, lumber, knocked-over trees, and dead bodies. There is so much water moving so fast that it will cause the collapse of many non-hardened structures. All of the debris and water will first flow inland; and then, when the wave is spent, much of the water and debris will flow back out to sea. Most human beings unfortunate enough to be on the ground or in a non-hardened building in that area get crushed or drown in the process. This is how more than 150,000 people lost their lives in just a few minutes in Banda Aceh. And any event that can kill so many people so quickly is definitely a doomsday scenario.

SCENARIO

There are now two types of tsunami-related scenarios that humanity has to worry about: earthquake-caused, or natural, tsunamis like the one described above and man-made tsunamis.

Natural tsunamis represent one of nature's greatest common threats to human life and property. They can occur around fault lines any time there is a major earthquake that forces a tectonic plate upward under the ocean.

Tsunamis have occurred for eons—as long as there have been earthquakes in oceans, there has been the potential for tsunamis to form. It is easy for geologists to find evidence of ancient tsunamis around the world. The first tsunami that humans recorded occurred in 479 BCE. Along Japan's coastlines, there are stone tablets, some as tall as 10 feet (3 m), inscribed with warnings such as "Remember the calamity of the great tsunamis. Do not build any homes below this point." These stones were erected over a century ago in response to a pair of tsunamis in Japan that killed tens of thousands of people. Those who experienced these tsunamis were so terrified by them that they felt it necessary to warn future generations.

The thing that is so frightening about tsunamis is the power of the water they contain. Moving water that is only 1 foot deep (30 cm) can shift a car. Just a few feet deeper, it can shift a house off its foundation. Even shallow water can easily knock someone over if it is moving fast enough. The problem with a big tsunami is that the incoming wave of moving water might be 20 or even 30 feet (6.1 or 9.1 m) high.

As if natural tsunamis were not bad enough, we now have to worry about man-made tsunamis as well. In 2018, Russia announced its decision to build an autonomous nuclear tsunami bomb that can freely roam the Earth's oceans. Such a bomb can threaten nearly any coastal city with complete annihilation via a gigantic tsunami wave. This relatively small, nuclear-powered submarine measures roughly 6 feet (1.8 m) in diameter and 65 feet (19.8 m) in length. Without humans onboard, there is no need to carry supplies, oxygen, or food. Therefore, it never has to surface. With a reliable nuclear power plant, the sub can theoretically run underwater for decades. Inside, the sub carries a compact, powerful nuclear warhead, armed and ready to detonate (see Nuclear Bombs, page 22). Positioning this submarine near a coastal city and detonating the nuclear warhead can create a tsunami wave of truly epic proportions and send it smashing into the city.

The bomb-induced tsunami could be hundreds of feet tall, capable of destroying everything in its path, including skyscrapers and other hardened structures. By tuning the weapon properly, salting the weapon for maximized radioactivity, and increasing the yield of the warhead, it can do even more damage by leaving behind nuclear contamination at a continental level. The hypothesis is that a salted weapon contaminates the water in the wave and the enormity of the wave carries it so far inland that a whole region might be contaminated for extended periods of time.

Either of these scenarios can be devastating to coastal regions. For example, Japan, which gets hit with natural tsunamis quite regularly because it is located near the Ring of Fire (see page 139), is especially vulnerable to this doomsday scenario.

LEFT This tsunami stone in Iwate prefecture in Japan warns its readers not to build on the land below this marker. **RIGHT** This pair of satellite photos shows the effects of the Indian Ocean tsunami. The top photo shows one area of the city before the tsunami arrived. The bottom photo shows the wreckage and the flooding that it left behind.

Tsunamis are an unfortunate reality on a planet like Earth, which is covered by fifteen major tectonic plates and many other minor ones.

Earthquakes tend to occur most commonly along the edges of these different plates, where they rub together or flow over and under each other when they move (see page 146). If the epicenter of a powerful earthquake is under an ocean and if the earthquake happens to force a tectonic plate upward, the probability of tsunamis becomes very high. When a plate snaps upward, all of the water in the ocean above it snaps upward as well. Tsunamis are the catastrophic waves that result from all this displaced water hitting a shore. Underwater landslides and volcanoes have also been known to cause tsunamis.

Since the lifted water spreads out in all directions, it means that tsunamis often occur in groups. A single earthquake can cause tsunamis to hit many different areas. Banda Aceh happened to be the closest major city to the epicenter of the 2004 Indian Ocean earthquake, so it only took about 20 minutes for the tsunami waves to roll ashore there. Many other shores were exposed to the danger, with the waves sometimes taking hours to arrive. Even India, hundreds of miles away from the epicenter, experienced over 10,000 deaths and 650,000 displacements, where people lost their homes to the waves.

TWENTY-FIRST-CENTURY TSUNAMIS

The set of tsunamis on December 26, 2004, was the most destructive tsunami event in modern history. Over 200,000 people died in the various tsunami waves that occurred around the Indian Ocean, and the total damage estimate is about $15 billion. But tsunamis are fairly regular events on our planet. Here are the three biggest tsunamis that have occurred just in the twenty-first century:

- **North Pacific Coast, Japan, March 11, 2011:** A magnitude 9.0 earthquake spawned this tsunami, which is also famous for causing the Fukushima Daiichi nuclear power plant disaster. Besides the tens of thousands of people killed by the waves, along with the widely publicized damage and aftermath at the Fukushima nuclear power plant, this tsunami destroyed or hobbled a great deal of the planet's production capacity for certain computer components, especially computer memory chips. More than 18,000 people died and 100,000+ buildings collapsed because of the earthquake and tsunami. It took more than a year for things to start getting back to normal.

- **Coast of Chile, February 28, 2010:** Two million people lost their homes in this tsunami and earthquake, and about 500 people died. The magnitude 8.8 earthquake that generated this tsunami was located just off the coast of Chile. Chile took the brunt of the destruction, but tsunami waves from this event made it as far as Japan and caused millions of dollars of damage even there.

- **Samoa, September 30, 2009:** This tsunami was caused by the most intense earthquake on the planet in 2009 at 8.1 magnitude, leading to hundreds of deaths. One wave that struck Samoa was 46 feet (14 m) tall.

When a tsunami wave hits the shore, it can be a massive, fast-moving event. Then, once the wave has spent itself, it recedes back toward the ocean, often carrying a load of debris with it. The speed and intensity of recession depends on the affected area's topography, because the receding water's motion is all driven by gravity.

PREVENTION

During the 2004 Indian Ocean tsunamis, there was no warning system to speak of in place. Some argue that the earthquake itself is a warning, but not all earthquakes result in tsunamis, and tsunami waves can travel and hit shores that are hundreds of miles away from an earthquake's epicenter.

Given this, there are two main ways for humanity to approach the threat of tsunamis. The first is to focus on the preservation of human lives. This approach requires three precautions:

- A warning system that lets people know that a tsunami is on the way
- An understanding in the affected population that evacuation from low-lying areas must occur on foot immediately and rapidly
- An evacuation plan that provides places for people to go.

A developing country like Indonesia, where funds are scarce and education is not universal, can have trouble developing and deploying a tsunami-warning system. In addition, even when a system is put in place, a nation can have trouble maintaining it or getting people to react effectively. For example, after the tsunamis on December 26, 2004, Indonesia received grants and technical assistance and installed 22 sensors on the ocean floor with buoys to facilitate the transmission of the sensors' information. Unfortunately, all the sensors were inoperable by the next time they were needed. They had either been damaged by natural forces, vandalized, or became inoperable due to lack of routine maintenance. In another incident in 2018, an earthquake caused immediate power failures, so any sirens that might be used to warn citizens about tsunamis could not sound. Many hundreds of people died.

Advance warning, whether from a warning system or even from the earthquake itself, can only be effective if everyone understands the danger of a tsunami and the proper response. This is why the population must be educated about tsunamis and evacuation routes. The ultimate form of preparation would be to hold regular "tsunami drills" for a city. Another step could involve extensive signage that warns people about the tsunami threat and points toward higher ground. The signs would raise awareness, while programs in schools could get kids on board.

With a warning system and some minutes available for people to flee, an evacuation plan can help save lives. People have to do one of two things:

1 They can move immediately to higher ground, in the form of nearby hills, for example, if there is higher ground available.

2 If there is no higher ground nearby, then people can move away from the shoreline, as tsunami waves typically do not travel farther than 2 miles inland.

An average person in good health can walk 2 miles away from a shoreline in 45 minutes, as long as they know the path. Under duress, a person in good health could cut that time in half. In a place like Banda Aceh, which was immediately adjacent to the earthquake's epicenter and offered only 20 minutes between the earthquake and the arrival of the tsunami, people would have needed to act instantly to save themselves. And thousands of lives could have been saved in places farther from the epicenter if people had known what to do and where to go, thanks to warning systems, education, and training.

Once lives are secured, the next approach focuses on the preservation of property. Japan provides an example of what is possible. To preserve property, Japan had built massive sea walls designed to block oncoming tsunami waves. The country's 2011 tsunamis were so enormous, however, that they overwhelmed and breached many of these sea walls. Therefore, Japan is redoubling its efforts, investing billions of dollars in even stronger, taller walls along many miles of shoreline. The new seawalls are as high as 40 feet (12 m) and must be designed essentially like enormous concrete dams. At the moment of the tsunami's peak, the seawall would hold back a lake of water up to 40 feet deep.

Individual high-value properties, such as power plants and factories, can be protected by their own seawall defenses as well. These defenses would surround each facility and create a protective seawall bubble. These kinds of protective measures are expensive, but less expensive than the loss of the facilities. Individual buildings like houses could be constructed in such a way that they could resist a tsunami. Whereas a wooden structure might be swept away by the force of a tsunami wave, a hardened concrete house could likely withstand it. However, a hardened house would be much more expensive, and what's inside would likely be flooded anyway.

Following the 2004 tsunamis, some countries have taken active steps to develop evacuation plans, evacuation routes, and evacuation areas. They have put up signage, designated specific safe places to go, and attempted to educate their citizens. A good example of planning can be seen in the state of Oregon. Here there is an imminent threat of tsunamis because of the Cascadia Subduction Zone off the Pacific coast. There will only be a few minutes

BELOW Workers close a gate in a seawall designed to protect Numazu, Japan.

for people to evacuate, and they will have to do so on foot. In response, the state has developed evacuation maps for coastal towns called "Beat the Waves." The basic goal is for people to run to high ground in the few minutes available before the waves hit. Maps (both printed and distributed as well as on signs) show where the high ground is and the quickest path to get there. Obviously, it would be best to look at the map well ahead of time, so you are not having to read one at the time of the tsunami.

The relative infrequency of tsunamis in many areas can lull people into a false sense of security or apathy. Structural fires are similarly infrequent, yet schoolchildren, college students, and employees in office buildings participate in annual fire drills to keep them prepared for the worst-case scenario. In the case of tsunamis, this same kind of vigilance is important for any low-lying coastal area where earthquakes can occur.

ABOVE Signage like this can help people understand what to do to avoid death by tsunami.

PREVENTING MAN-MADE TSUNAMIS

If Russia really does have autonomous nuclear tsunami subs roaming the world's oceans, what can the world to do about it? One deterrent is the decades-old principle of mutually assured destruction, where the world promises to annihilate Russia if it ever uses its new tsunami weapon.

The other alternative is to ban all military submarines. This would not be easy, but it would be possible with a concerted worldwide effort. Why might we ban military subs? Because the idea that any country has invisible submarines carrying city-killing nuclear missiles roaming freely all over the planet is fantastically dangerous. A military submarine's intent is purely destructive.

In a world where destructive weapons are hard to ban, how do we eliminate the threat of tsunami-generating submarines? We need global sensor networks that can detect them underwater by using acoustic (sonar), magnetic, or thermal sensing techniques. There are many creative robotic detection systems that could be deployed. Imagine, for example, a million solar-powered robot ships that are constantly scanning the ocean beneath them for subs. When any sub is detected, it can be destroyed.

This fleet of robot ships could also solve many other problems. They could detect illegal fishing, find modern pirate ships, discover drug runners and slave runners, detect illegal releases of waste and chemicals, and more. In addition, they could be packed with sensors to conduct important scientific research on the global oceans.

CORONAL MASS EJECTIONS

THREAT LEVEL: COUNTRY

As soon as NASA made the announcement about the coronal mass ejection, every news organization on the planet started running wall-to-wall doomsday stories about it. The essence of all these stories was nearly identical: the United States is doomed because a massive "bomb" of solar particles that blasted out of the sun will be arriving in a few days. Most scientists predict that the power grid will collapse. There may not be electricity for months, maybe a year or more. And this will plunge the country into chaos.

An explosion on the surface of the sun caused this coronal mass ejection, or CME, and it has released 100 million tons of charged particles. The particles are mostly protons, neutrons, and helium atoms stripped of their electrons. These particles travel through the vacuum of space at a speed of more than a million miles per hour (1.6 million km/h). Because Earth is roughly 93 million miles (150 million km) from the sun, it takes a few days for the particles to arrive and hit the planet. Scientists have used this information to calculate which side of the planet will be facing the sun, and it looks like the particles from the CME will make a direct hit on the United States.

When these 100 million tons of charged particles hit the planet, we are told, it won't be a good day for human beings in the modern world. How much is 100 million tons? Imagine your typical Olympic-size swimming pool, which measures 50 × 25 × 2 meters (164 × 82 × 6 feet). The pool contains about 5 million pounds (2,300,000 kg) of water. One hundred million tons of water would fill 40,000 of these swimming pools. When this amount of charged particles travels to Earth at more than a million miles per hour, they are going to cause a big effect. Most importantly, they will disrupt the power grid.

The particles will interact with the Earth's magnetic field and atmosphere, causing the magnetic field to distort and then reconnect. In the process, the country's long transmission lines will pick up massive amounts of current and send it to transformers and other parts of the system. These surges of current will overload all kinds of equipment, causing widespread failures and taking down the grid. With the power down for months, daily life will come to a standstill. Drivers can't pump gas, and lights won't work. Nor will refrigeration, heating, and air conditioning. How will food and medicine be delivered to stores if there is no power? How will the stores be open if there is no power? How will the factories make food and medicine, or refine gasoline, or do anything else if there is no power?

RIGHT NASA's Solar Dynamics Observatory captured this photo of a coronal mass ejection in June 2015.

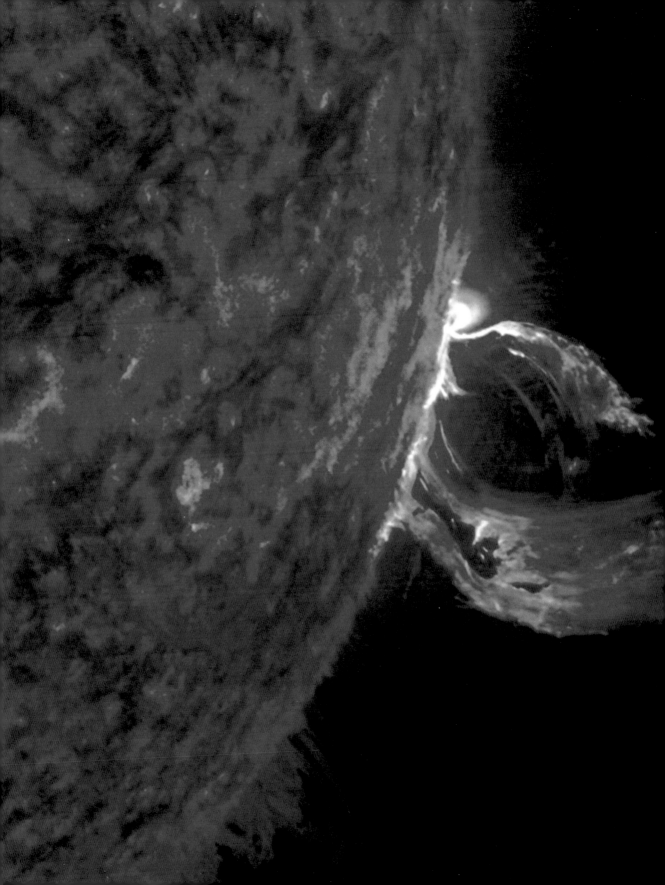

Though the country's government and power companies have been anticipating this kind of CME event for decades, and despite humanity's knowledge about CMEs dating back at least 150 years, officials seem to be completely and infuriatingly unprepared. As we hear story after story about how bad the effects could be, some hold out hope that the reports are exaggerated. The only way to see the reality of the situation is to wait for it to hit. If the effects are anything like the news stories and predictions are suggesting, we are all in big trouble.

SCENARIO

CMEs are common. The number that happen on any given day will vary depending on where the sun is in its 11-year activity cycle. When the sun's activity is low, there might be one CME a week. When solar activity is high, there might be five CMEs a day.

If CMEs are this common, then why isn't the Earth bombarded all the time? One reason: the sun is a sphere, and this shape makes the probability of a CME being aimed exactly in Earth's direction low. Another reason: CMEs vary in size and intensity, and smaller ones don't have much of an effect.

Nonetheless, CMEs have affected the Earth more than once throughout human history. The most famous CME occurred in 1859 and is known as the Carrington Event. Though 1859 predates power grids and telephone service, the first telegraph wire of any significant distance was set up in 1844, bridging the 40 miles (64 km) between Baltimore and Washington, DC. By 1859, there were many telegraph wires crisscrossing the country. The two-day Carrington Event was so powerful that it set some telegraph wires on fire and caused general disruption to many telegraph systems. It also caused auroras, which we normally associate with the polar regions, to stretch all the way down toward the equator. These auroras were so bright that at night, people could read by their glow.

If a major, direct-hit kind of CME like the Carrington Event occurred in the modern era, the consequences could be dire. A report from the Department of Energy called "High-Impact, Low-Frequency Event Risk to the North American Bulk Power System" includes this warning:

> Geomagnetically-induced currents on system infrastructure have the potential to result in widespread tripping of key transmission lines and irreversible physical damage to large transformers . . . . The physical damage of certain system components (e.g. extra-high-voltage transformers) on a large scale, as could be effected by any of these threats, could result in prolonged outages as procurement cycles for these components range from months to years.

Damage estimates in a worst-case scenario range into the trillions of dollars—a result of both repair costs and lost productivity. If a CME were to hit at a critical point in time, such as in the afternoon on a hot summer day when transmission lines would be humming at maximum capacity, it could overload the lines, along with the related transformers, and lead to an even bigger mess. Significant transformer damage, and other system damage, could cause months to elapse before power could be completely restored.

The big question here is this: What would cause the sun to eject 100-million-ton blasts of charged particles out into space? After all, the sun started as a big ball of hydrogen gas. How does a big ball of hydrogen become so violent?

The sun is utterly enormous—it has the same volume as about 1,000 Jupiters or 1.3 million Earths. The sun's immense size is key here. It contains so much hydrogen in one place, and so much gravity from the sheer mass of all this hydrogen, that the compression forces inside the sun are enormous. There is enough gravitational compression to crank up a fusion reactor in the center of the sun, forcing the hydrogen atoms in the sun's core to fuse together into helium atoms.

You would think that the energy of all these fusion reactions would blow the sun apart, just as a fusion bomb (see page 30) blows apart. However, the intense gravity created by the sun's size keeps it intact. The enormity of the sun means that it is fusing something like 50 trillion tons of hydrogen (45 quadrillion kg) per day to create its heat and light. That sounds like a lot, but since the sun weighs something like 2,200 trillion trillion tons, it is hardly even a drop in the bucket.

When two atoms of hydrogen fuse and create an atom of helium, this new helium atom has a slightly lower mass than the total of the original two hydrogen atoms. This difference in mass converts into energy, primarily in the form of heat and light. (This is a significant simplification of what actually happens, but it provides a nice summary of the process.) The sun produces so much heat and light that we can easily feel it here on Earth, 93 million miles (150 million km) away.

ABOVE An aurora glows in the night sky above Whitehorse, Canada. The aurora appeared as a result of a CME that occurred three days earlier in October 2012.

MAJOR CMES THROUGHOUT HISTORY

Since 1859's Carrington Event, there have been dozens of other events of varying intensity, including:

- **1877:** This CME was reported in the *New York Times* because of the auroras it caused. The report noted that "not since the wonderful exhibition of the Spring of 1869 has anything so sublime been observed as the play last night of fire that glowed vividly from horizon to zenith for more than two hours."
- **1882:** This event led to more problems for telegraph operators.
- **1903:** By this time, there was a telegraph cable crossing the Atlantic Ocean. The CME disrupted its normal operations.
- **1940:** With telephone systems now more common and existing alongside telegraphs, CMEs begin to affect more telecommunication systems.
- **1989:** Thanks to this CME, Quebec experienced a power-grid failure that nearly spilled over to the eastern US grid. This power failure affected the entirety of Quebec and lasted 12 hours. This CME was much smaller than the 1859 event, but even so, it produced auroras that were seen as far south as Cuba.
- **2003:** Solar flare activity spiked high enough to affect many satellites. The Japanese satellite named ADEOS-II was completely destroyed in this storm.
- **2005:** Another solar flare disabled GPS satellites for several minutes.
- **2012:** A storm with an intensity similar to the Carrington Event occurred and crossed Earth's orbit, but it fortunately was a near-miss rather than a direct hit.

CMEs happened before 1859, but because there weren't any electronic devices or telecommunication systems, no one paid much attention. The Carrington Event was the first major solar storm that could interact with anything that used long-distance lines and electronics on Earth. As our technology gets more and more advanced, our vulnerability to solar storms increases with each passing year.

This gigantic-ball-of-hydrogen-with-a-fusion-reactor-at-its-core idea is weird enough. But then things start to get even weirder. The sun starts to create its own internal "weather systems." The three most common phenomena that occur on the sun are sunspots, solar flares, and coronal mass ejections.

These happen because of the sun's strong magnetic field. This magnetic field appears because the atoms inside the sun are so hot that they separate from their electrons and become plasma. The negatively charged free electrons and the positively charged electron-less atoms move around due to convection, creating currents and thus magnetic fields. The interactions between the plasma and the magnetic field lead to things like solar flares.

A solar flare creates a purely electromagnetic effect by shooting electromagnetic waves into space. This means that when there is a solar flare that is pointed toward Earth, we feel the effects 8 minutes later, as the electromagnetic waves travel at the speed of light and we are 8 light-minutes from the sun. With a solar flare, the Earth gets hit primarily with X-rays and gamma rays. Because of the Earth's atmosphere and its own magnetic field, those of us on the ground feel no ill effects from a solar flare, although it might hamper some radio broadcasts or affect satellites in orbit. An astronaut on the moon, however, where there is no atmosphere for protection, would get a solid yet non-fatal dose of X-rays from the typical solar flare.

A coronal mass ejection is a completely different beast. In this case, an explosion occurs on the surface of the sun that shoots millions of tons of matter—as described earlier—into space. The matter primarily consists of plasma: hydrogen and helium atoms stripped of their electrons. Again, Earth's thick atmosphere protects us. The particles do not affect us biologically, so living things on Earth are okay. But the charged particles interact with the Earth's magnetic field to create geomagnetic disturbances (GMDs). The GMDs create geomagnetically induced currents (GICs) that threaten the power grid by inducing massive destructive currents in long wires. Preventing these GICs from taking down the grid and damaging its important components is the key to surviving a CME event.

ABOVE In this engraving from 1873, a repairman climbs a telegraph line.

PREVENTION

Preventing the destructive effects of a CME on the power grid involves tactics that are very similar to those used to mitigate the effects of an EMP attack (see page 44) or a grid attack (see page 90). The first step would involve completely hardening the power grid so it can shrug off CMEs. The basic idea is to add equipment to each large transformer that protects it (along with some other similar adjustments grid-wide). The transformer equipment is expensive (one estimate is $350,000 per transformer), and there are on the order of 100,000 large transformers across the US grid alone. It will take time and perhaps $30 billion to get the job done. $30 billion may sound like a lot of money but not in comparison to the cost of recovering from an unprotected large-scale CME event.

NASA has an interim program in motion, called Solar Shield, that is trying to identify and protect

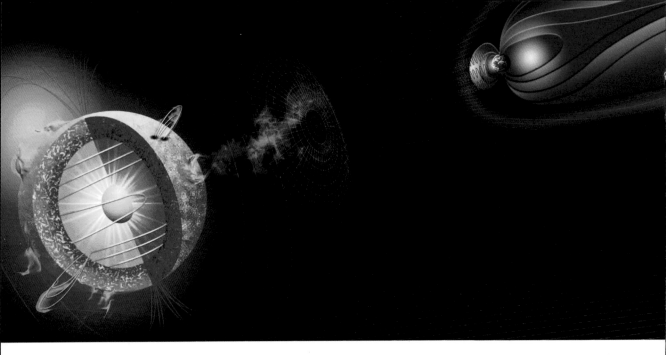

ABOVE Particles from a CME can distort Earth's magnetic field. As the field distorts and snaps back, the changes induce currents in power lines that can bring down the grid.

high-risk transformers. The protection comes in the form of a system that can predict when a CME storm will arrive. This allows us to disconnect high-value transformers (and other components) from the grid temporarily before the storm rages. With the transformers disconnected, the massive currents that develop in the long transmission lines cannot impact the transformers. In other words, we take the grid down on purpose rather than letting a CME destroy the grid.

This approach is not completely foolproof, nor is it easy without rehearsals. Imagine what would happen if one day, out of the blue, FEMA suddenly announced that within twelve hours, the entire nation's power grid will go down for two days as a preventative measure for an incoming CME. Panic and dysfunction would likely ensue, absent societal and technical preparation. In other words, it would be wise to prepare people for a CME grid shutdown ahead of time with something like a power-grid drill. If we were all prepared in this way, the good

news is that the act of taking down the grid and isolating components for a couple of days would help protect the lines, the transformers, and the billions of devices that are connected to the grid. We would have a much better chance of coming out of a CME event intact, with only minor damage.

What if we wanted to make the world, which is highly dependent on the grid, even more resilient in the face of a CME event? What can individual households do to increase their preparedness further? One approach would be to come up with ways to maintain communications and internet access in all households and businesses across the country regardless of the power grid's status. Doing this would allow everyone in the country to stay informed and communicate freely during the time the grid is preemptively shut down. These steps would reduce panic and boredom during the event and allow important information to keep flowing. In other words, it would be beneficial if each home could be energy-independent for two days if there is a CME event where the grid is down. First, each home would need to have enough battery capacity available to power the household's internet equipment. This equipment includes the home's internet modem and router, along with laptops and phones used to access the internet. Because these devices do not draw a tremendous amount of power, a two-kilowatt-hour battery would be able to carry the load of most households. This battery system might cost between $500 and $1,000 at current prices, probably less with mass production and distribution. The knee-jerk reaction to this idea might be to say it is too expensive. But consider the fact that just about every American home contains a refrigerator in the same price range and a furnace or HVAC system costing five times more. If the power grid is going to be down for two days, the second step involves ensuring that people have access to fuel for generators and transportation. This means that the entire fuel system, including gasoline pipelines, storage and distribution systems, and gas stations, needs to be able to operate without a power grid if

necessary. How is this possible? The network could be run on the fuel that it carries. Gasoline generators could power the system's components, allowing them to run independently during a temporary power outage. Same for diesel fuel and jet fuel networks—these fuels can power generators to keep the pipelines and distribution systems running.

With all these measures in place, it would be possible to avoid major damage to a country's power grid by taking it down for a few days. Homes might not be supercomfortable without air conditioning, and the food in our freezers might melt, but we could all access the internet and drive if we needed to go somewhere in an emergency. When the CME event finishes, we would power the grid back up and be back in business. These measures would have a cost, just like the hardening of all transformers has a cost. But the reward in terms of preparedness might be well worth it and create a highly resilient society that could easily handle this kind of short-term catastrophe.

BELOW To prevent a CME from causing large-scale damage, the power grid can be taken down in advance.

MASS EXTINCTION

THREAT LEVEL: WORLD

When we think of doomsday scenarios, we usually view them through a human-centric lens and think of events that have a big negative impact on our civilization. For example, we see hurricanes, earthquakes, and tsunamis as doomsday scenarios for humans because they destroy major cities and can kill a significant number of people. But what if there is a doomsday scenario that is already under way on Earth, one that does not primarily impact humans? What if, instead, millions of other species on Earth are facing one of the largest, most destructive doomsday scenarios possible?

The scenario that we'll discuss in this chapter is known as Earth's sixth mass extinction event. There have been five mass extinction events so far in the planet's history—all of them caused by nature in one way or another. This sixth event is different. In this case, humans are acting together to destroy half, or more, of the life on this planet. Millions of species will become extinct because of human activities if we continue current trends. Quadrillions, perhaps quintillions, of living beings may perish. Based on these numbers, the sixth mass extinction event will be the greatest doomsday scenario ever.

To understand how this doomsday scenario is unfolding right before our very eyes, let's take a look at the fish in the ocean.

For thousands of years, human beings have harvested fish primarily for food. Human fishing has occurred at a small scale for many millennia, and it largely took place at a sustainable pace. Even by 1800, the global human population was only 1 billion. Because humans had not yet invented the tools used by today's industrial fishing fleets and widespread ocean pollution was uncommon, fishing's impact on the oceans was tiny. Humans could only kill so many fish with their relatively small human-powered or wind-powered boats, and fish could easily reproduce and recover. Life for fish in the ocean was pretty good.

But things started to change during the nineteenth century. Whales saw these changes first. Whales are not fish, of course, but these ocean-dwelling mammals are a canary-in-the-coal-mine of sorts for fish-kind. Many whales can't swim very fast, and they have an Achilles heel—they need to come to the surface every so often to breathe. Even in

RIGHT A fishing boat uses a purse seine net to catch herring. One catch can capture hundreds of tons of fish.

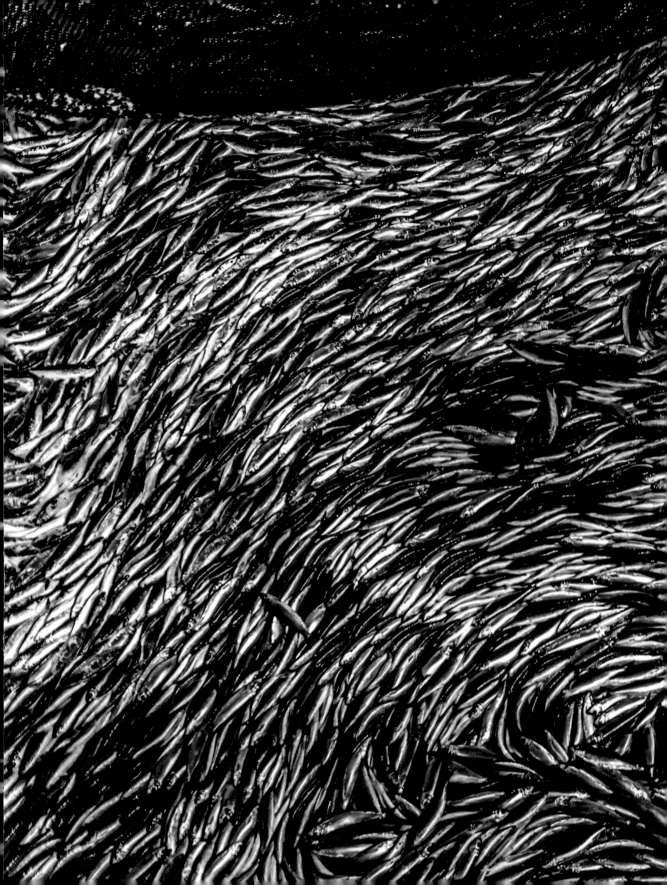

the 1800s, whales were in danger from the whaling industry. A dead sperm whale contains tons of blubber and oil worth a huge amount of money, making them a lucrative target for an industry that grew rapidly to exploit them and other whale species. Once factory ships got rolling in the 1920s, allowing dead whales to be dragged right onto the deck for quick and complete dismemberment and processing, whale populations plummeted. By the 1960s, many species of whales were practically extinct due to the relentless slaughter. Blue whale populations fell as low as 5,000 individuals. The only thing that saved whales from complete extinction was a near-worldwide moratorium on commercial whaling by the International Whaling Commission (IWC).

By the twentieth century, the fishing industry had started to change similarly with the invention of steel ships, big diesel engines to propel them, and powerful electric motors to pull in nets. These new technologies made it easy to deploy and haul in huge purse-seine nets and trawling nets. Now a few humans could cast these gigantic nets and catch millions of small fish, or thousands of large fish, at once. Today, diesel engines, electric motors, steel ships, huge drag nets, underwater sonar to find fish, aerial fish spotting, GPS systems, and other technologies allow fishermen operating huge factory ships to catch tons of fish at a time.

As the population of humans grows, so does the demand for fish. Therefore, the incentives to catch more fish become stronger every year. The problem with the scale of industrial fishing today is that it overwhelms the oceans, putting massive pressure on fish populations and threatening to collapse the world's fisheries. Right now, humans are extracting 90 million tons of fish from the ocean every year, not including undocumented fishing, illegal fishing, and bycatch (fish that are caught "accidentally" and therefore thrown away). Some of these fish species are tiny, like sardines, and some of them weigh hundreds of pounds, like tuna. It is thought that fishing fleets will have extracted just about every fish from the ocean by 2050. These fishing fleets do not care about the future. They care about how much money they can make today. It is the tragedy of the commons, writ as large as possible. Combine overfishing with human waste, plastics in the ocean, oil spills, pesticides, and climate change, and it is a recipe for disaster.

What does it look like when overfishing occurs? The northeastern United States fishery provides one example. Cod were once so plentiful that people would talk about walking across the water on their backs. There is even a place named for the fish—Cape Cod, Massachusetts. After the arrival of giant factory ships with huge nets, the cod fishery collapsed.

Another example of a collapse that is easy to see is the oyster population in the Chesapeake Bay. Oysters were once so plentiful in the bay that oyster reefs could endanger approaching ships. It is thought that there were so many oysters filtering the water of the bay that they could filter all 18 trillion gallons (68 trillion l) there in just one week. But because of overfishing and pollution, the number of oysters has plummeted; only about 1 percent of the original population is left in Chesapeake Bay.

Several species of tuna are also at risk of collapse. According to the National Oceanic and Atmospheric Administration's (NOAA) assessment of the Pacific bluefin tuna in 2016, "the spawning stock biomass was at 3.3 percent of the level it would be had the stock never been fished." In other words, if we could go back to 1940 and compare that to today, 97 percent of the Pacific bluefin tuna are gone.

There is no question that human activities have the ability to completely destroy entire species.

We have focused on fish in the discussion above because we have detailed statistics about fish. In many developed nations, fish harvests are tracked and regulated to some degree. Therefore, we know, with certainty, that many species of fish are in danger of becoming extinct unless humans chart a far different course going forward.

From a scientific standpoint, fish are just one slice of a much larger pie. There are thought to be 8.7 million different species on the planet, and millions of these species are heading toward extinction. How are humans driving the Earth's sixth mass extinction event? We release gigatons of CO_2 and methane into the atmosphere through fossil-fuel combustion, livestock, thawing permafrost, and deforestation. These activities cause global warming (see page 52), altering the habitats of many species, along with flooding and drought from climate change. Pollution from oil spills, runoff, human waste, smog, plastics, and much more, along with human development, have also caused loss of habitat. The use of pesticides and the side effects of this pesticide use have made certain species particularly vulnerable. For marine life, ocean acidification (see page 200) and pollution have wreaked havoc. Even the lights we use at night are having an impact on animals, especially insects. Insects that are attracted to lights die instead of living out their natural lifecycles. As one species dies off, those that eat them are also threatened, potentially upending entire ecosystems.

In the past fifty years, the number of extinctions has grown exponentially. We have already seen how fish-kind, along with all other marine animals, are heading toward extinction, but other types of animals will disappear in similar ways:

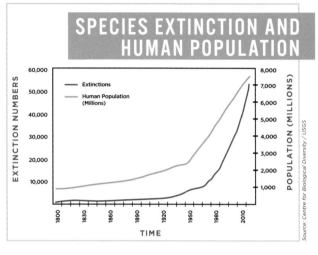

SPECIES EXTINCTION AND HUMAN POPULATION

Source: Centre for Biological Diversity / USGS

ABOVE Extinctions have grown exponentially alongside the human population.

■ **Insect-kind faces threats, mostly from pesticide use and loss of habitat.** Researchers have found that if we weighed all the insects alive on the planet today and then in a year we weighed them again, the total weight would decrease by 2.5 percent per year. At this rate, insects will disappear from the planet in a few decades.

■ **Bird-kind is struggling due to loss of habitat as well as loss of food sources, disruption of migratory routes, predators such as cats, and human-built structures including wind turbines.** In the last fifty years, North America has lost a third of its birds. If there were 9 billion birds flying around North America in 1970, there are 6 billion flying around now, and the number keeps falling. Pesticides kill birds. Cats kill birds. Laws meant to save birds are being rolled back. Eventually, there will be no birds left in North America either.

- **Reptile-kind is under pressure due to factors ranging from unregulated collecting to climate change.** The International Union for Conservation of Nature has pointed out that 20 percent of reptile species on the planet are already either endangered or in a state of near-extinction.

- **Amphibian-kind is in danger because they are especially sensitive to pollution and climate change.** As many as one third of amphibian species are already endangered.

- **Many mammal species are under pressure, and many have gone extinct recently.** The West African black rhino became extinct in the last decade, mostly because of poaching. The Formosan clouded leopard met its demise due to habitat loss and poaching, while the Caribbean monk seal was hunted for meat and for being a predator (much like wolves). The Christmas Island pipistrelle, a small bat, is now extinct because of pesticides. The Madagascar hippopotamus is extinct due to hunting. The Yangtze River dolphin succumbed to both water and noise pollution. Even iconic species are declining to the point where they will soon be extinct in the wild. The African elephant, for example, is being driven toward extinction by hunting, poaching, and habitat loss. There are only 4,000 tigers left in the wild.

Why does this gigantic planet-wide extinction process matter to humans? All living things are connected. If humans are making Planet Earth so hostile to life that millions of species will perish, it is safe to assume that we will also imperil the species we need. Plants and animals that all humans depend on for food are going to feel the same pressures that wildlife is feeling. In the case of fish, 3 billion people in the developing world—approximately half of the people in the human species—live near the ocean and depend on fish that they can catch in order to eat. For 1 billion of these people, fish represent half of the protein they eat. Once we eliminate all the fish, these people face starvation.

A planet that becomes this inhospitable to wildlife is quite likely to become hostile to humans as well. Humans are, in essence, preparing to destroy themselves in the end.

SCIENCE

Officially, a mass extinction event is a catastrophic episode in the planet's history where lots of the multicellular life has perished in a relatively short interval of time.

The universe is estimated to be 13.77 billion years old. Planet Earth is estimated to be 4.5 billion years old, formed (like the rest of our solar system) out of the supernova remnants of another star. Once Earth formed and stabilized, when did life on our planet begin? To answer this question, we look to the fossil record. Keep in mind that any fossil is, by definition, made from the remains of a living thing. The oldest fossils paleontologists have

RIGHT Factors ranging from habitat loss to climate change have threatened or endangered the population of these animals, along with many other living creatures. Clockwise from top left: a koala bear, common loon, honeybee, gold dust day gecko, and green sea turtle.

WHAT IS A GEOLOGICAL TIMESCALE?

When human beings think about time, we tend to think in terms of things like minutes, days, years, and centuries. Something that is "a *long* time ago" might be several thousand years in the past.

But when geologists think about time, they think in terms of geological events. They think in terms of eons, eras, periods, epochs, and ages. A million years is a short period of time for a geologist. When we think about the Jurassic period, made familiar by the movie *Jurassic Park*, we are talking about a period roughly 56 million years long. It started about 200 million years ago and ended about 150 million years ago. Two hundred million years ago, the supercontinent Pangaea started dividing into two large continents, marking the beginning of the Jurassic period. There was also a mass extinction event at the beginning of the Jurassic period.

found indicate that life is 3.5 billion years old. Life on Earth might be older than this, but we have not yet found anything in the fossil record to prove it.

Life on Earth started with just one species of simple bacteria that arose spontaneously. These ancient life-forms were very simple, the predecessors to the bacteria we find on Earth today. They did not use oxygen to survive, because Earth's atmosphere had no oxygen at the time.

The entire tree of life, all of the species we see today, started with just one cell. Since this starting point, a billion or more species have evolved over time. Ninety-nine percent of them are now extinct, because extinction is a completely natural phenomenon. Species go extinct for a variety of reasons, including competition, loss of habitat, and changing weather.

A mass extinction event happens when something sudden (in geological terms) and utterly catastrophic in scale happens. The event rapidly kills off a large number of species. It took about 3 billion years for the first mass extinction event to occur, but it only took 500 million more years for four additional mass extinctions to happen.

- **Ordovician-Silurian extinction, 445 million years ago:** This was the first of the five great mass extinction events. Perhaps 85 percent of all life on Earth perished then. The cause was a period of glaciation, followed by a large drop in sea level as more and more water was locked up in the ice. It is thought that new mountain ranges were forming, which took large amounts of carbon dioxide out of the air, leading to global cooling.

- **Late Devonian extinction, 360 million years ago:** This mass extinction event took longer and is less well defined than the others, and is attributed to climate change, perhaps driven by asteroid impacts and a lack of oxygen caused by colossal algal blooms. Seventy-five percent of species were lost.

- **Permian-Triassic extinction, 252 million years ago:** Known as the Great Dying, this mass extinction event which defines the end of the Permian period was a worst-case scenario, taking out something like 95 percent of the life on Earth. This one is also thought to have been spawned by volcanic activity.

AN UNCOUNTED EXTINCTION EVENT

Two and a half billion years ago, there was a notable extinction event that does not get "counted" because multi-cellular life did not yet exist. At the time, Earth's atmosphere was oxygen-free (similar to Mars and Venus today), so all of the life on Earth was anaerobic: it did not require oxygen for its metabolism.

Then blue-green algae, also known as cyanobacteria, evolved into existence. These early cyanobacteria are the single-cell species that first harnessed photosynthesis, and so they started pumping oxygen into the atmosphere. It took a long time, but eventually there was enough oxygen in the atmosphere to start killing off the anaerobic bacteria that had ruled the planet. Today, anaerobic bacteria exist only where they can find an oxygen-free area. But the good news is that the oxygen opened up a new niche for evolution to exploit. Thus, we have all of the oxygen-breathing bacteria, insects, and animals that we see today. The full evolutionary process unleashed by the oxygen unfolded over the course of billions of years.

- **Triassic-Jurassic extinction, 200 million years ago:** This mass extinction event destroyed about 75 percent of the species on Earth. This one is attributed to massive volcanic activity in the Atlantic Ocean and possibly an impact event as well.

- **Cretaceous-Paleogene extinction, 66 million years ago:** This mass extinction event is the best known out of the five. Asteroid strikes, some serious volcanic activity happening around the same time, and the resulting impact winter killed off not only the non-birdlike dinosaurs but approximately 75 percent of all species on Earth.

After a mass extinction event, it typically takes millions of years for life to recover to its former levels. First, the cause of the event must be resolved so the climate can restabilize. Then, the species that survived have to reestablish their populations. And then new species must evolve from them to fill all the newly empty niches in the ecosystem.

The mass extinction event that we are experiencing today has the potential to destroy so many species, and so much of the life on Earth, that it will again take millions of years to recover. Just think about what will potentially unfold. We could destroy most of the species that inhabit the ocean today, along with all the rainforest species around the equator, along with many of the insect, mammal, bird, reptile, and amphibian species that we see today. Why? In a worst-case scenario, the entire equatorial region becomes uninhabitable because of the heat. The ocean has been acidified so much that most marine species are gone. All of the ice has melted, so a great deal of habitable land is lost. Humans and the surviving life-forms are forced onto what land is left toward the poles. Millions and millions of species will be annihilated in the process.

BACKGROUND EXTINCTION RATE

Extinctions do naturally occur, but they are rare. For example, the "background" (natural; not human-caused) extinction rate for mammals is one mammalian species every 200 years. So in a million years—a geological kind of time frame— about 5,000 mammalian species would go extinct through natural causes. Today, however, extinctions are happening many times faster than the background extinction rate. In the last 100 years, dozens of mammal species have gone extinct, and many more are on the brink. If a major habitat like the Amazon rainforest collapses, a million total species could go extinct in a few decades. Because extinctions are happening now so much faster than the background extinction rate, we are in the middle of a mass extinction event, and the pace is accelerating.

BELOW Our fossil record shows the remains of millions of species that have perished from past extinction events, including many species of ammonite, a type of marine mollusk.

Today, human beings are overseeing the extinction of thousands of species. Soon, that number will accelerate as more and more species are affected. Eventually we are likely to kill millions of species off in this sixth mass extinction event—unless we change course. If we could somehow wave a magic wand and stop all of this death and destruction, what might that look like?

Let's look at Madagascar as a representation of the planet in miniature. Madagascar is an island nation off the coast of Africa that is about 1,000 miles (1,609 km) long and 230 miles (370 km) wide. It is roughly the size of Texas. If we could go back in time 10,000 years, the island was a tropical forest paradise filled with wildlife, free of humankind.

When humans arrived, they started exploiting the resources on the island, as humans are wont to do. Very soon, the humans were burning down the forests for farming and destroying habitats for many different species. Several species of megafauna, including giant lemurs and elephant birds (which weighed about 1,600 pounds, or 725 kilograms), were hunted to extinction for food.

Over the last fifty years, the human population of Madagascar has grown from 5 million to 25 million people, a five-fold increase. The ecosystem on the island is rapidly collapsing as a result. More humans mean more loss of habitat for wildlife, because the humans need space for housing, roads, and agriculture, among other needs. Activities like logging and mining also take a toll on the land. The status of lemurs offers an example of the effects that these activities have on wildlife. There are 111 species of lemurs left on Madagascar today, and all are expected to soon be extinct unless drastic action is

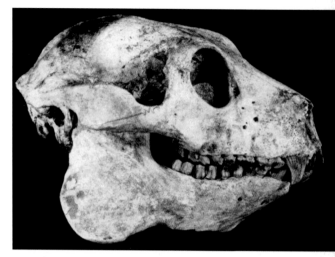

ABOVE A skull of a lemur that belongs to the genus *Babakotia*. Lemurs from this genus once lived in Madagascar but became extinct shortly after humans arrived.

taken to limit habitat loss. Today, humans also continue to hunt lemurs for food, and they also capture and export them for pets.

When does the destruction of wildlife on Madagascar stop? If we look back in history, one answer is that the destruction doesn't stop. Think about the island of Manhattan. With the exception of Central Park, Manhattan is an island covered in humans and human structures, with no wildlife at all. Most cityscapes that humans create end up like this. As the human population continues growing, this is the likely end point for Madagascar as well if nothing is done. Every unique species on the island will be annihilated.

What is happening on Madagascar can be seen as a miniature, higher-speed version of what is happening to Planet Earth as a whole. So what are we to do to avert catastrophe?

One magic-wand solution for Madagascar is to eliminate all humans from the island, remove human structures, and then let nature take over and repair the island. Why not let Madagascar return to being a tropical paradise? If not this, the next best thing would be to cut the island in half. Humans would live on one half and be completely excluded from the other. The unique Madagascar biome could be saved.

What if this idea of preserving half the island for wildlife were applied to the whole planet, in each area? Biologist E. O. Wilson advocates for this simple idea in his book *Half Earth*. As discussed in Rainforest Collapse (see page 178), human-free global parks protecting important ecosystems would allow nature to repair the land and its ecosystems and let threatened and endangered species flourish. These parks would help preserve the millions of species there that will be lost if the rainforests and boreal forests collapse. Where will the humans go? Wilson proposes greening the Sahara Desert and moving people there as one possible solution for Africa. The chapter on rainforest collapse offers a solution for preserving the Amazon rainforest and other forests. Cutting down on the human reproduction rate,

and eventually shrinking the human population, is another worthwhile approach to consider.

For the species that live in the oceans, we need to go further because they face so many threats. The solution involves immediately eliminating all industrial fishing before mass collapse of the ocean's fisheries can occur and providing funding to accelerate the rise of aquaculture. Aquaculture production has grown rapidly over the last thirty years. For example, salmon farming in Norway has increased from 100,000 tons per year to 1.3 million tons per year in that time. The amount of fish sourced from aquaculture now equals the amount of fish drawn from the oceans. By accelerating the growth of aquaculture, we can fully replace the amount produced by the world's industrial fishing fleets and then completely eliminate industrial fishing so the oceans can recover.

Aquaculture today is not perfect by any means, so we would need to address its environmental impacts. A comprehensive solution to pollution in the ocean, especially from developing nations and the oil industry, would also help fish populations recover. Ending fossil-fuel use would prevent any additional oil from contaminating the oceans, and eliminating single-use plastics (as many nations are starting to do) will help stop plastic pollution. In addition, we need to implement the solutions to reverse ocean acidification (see page 200).

We have seen this human-free approach work on a small scale before. The Korean Demilitarized Zone, a strip of land between North Korea and South Korea, is largely free of humans because of fences and treaties. Nature is thriving there. The 2,000-square-mile (5,180 km^2) exclusion zone that protects people from the radiation around Chernobyl's nuclear power plant has allowed nature to reclaim much of that area.

Human beings are the direct cause of Earth's sixth mass extinction event. Through climate change, pollution, habitat destruction, overconsumption, pesticide use, and many other offenses, human beings will destroy millions of species unless we take concrete and massive action to change course. The question is whether or not humans can come together to behave responsibly and avert one of the greatest doomsday scenarios ever faced by the planet. If we do not, our behavior as a species is unconscionable.

LEFT Established in 2001, the Anja Community Reserve is known for having the highest density of endangered ring-tailed lemurs in Madagascar.   **BELOW** The Chernobyl Exclusion Zone covers 1,000 square miles (2,590 km^2) and has become a sanctuary for many of the native species in that region.

RAINFOREST COLLAPSE

THREAT LEVEL: WORLD

As I stand in the desert that was once the Amazon rainforest, it is impossible to believe that humanity let this enormous tragedy happen, especially given what was at stake. This place where I am standing was once an impossibly lush jungle of trees, vines, plants, animals, and insects. Three million unique species lived here once upon a time. The area received over 100 inches (254 cm) of rain per year. It was beautiful. It was vibrant. It was as alive as it could possibly be.

Years ago, I was a scientist who was doing everything in my power to try to educate humanity on what was going to happen to the rainforest if we did not radically change course. And what I found was that I might as well have been talking to a brick wall. There were scientists, environmentalists, and concerned citizens who wanted to stop this tragedy we called rainforest collapse, but so many more were interested in the dollars they could gain from razing the land.

A few decades later, the collapse unfolded before our eyes, exactly as predicted. What was once the Amazon rainforest is now a desert. How did this catastrophe happen? Bringing up an analogy can help.

Imagine that we could travel back in time to the age of the Lewis and Clark expedition, which began in 1804. At that moment in history, the western United States (everything to the west of the Mississippi River) was largely untouched by European immigrants.

At that time, there were millions of acres of enormous old-growth redwood trees in the United States. There were two species: the giant sequoia (larger by volume) and the coast redwood (larger by height). These old-growth redwood trees are enormous, growing 300 feet (91.4 m) tall or more and 20 feet (6.1 m) or more across, and are often thousands of years old.

Ninety-six percent of these magnificent redwood trees are gone today. Nearly every single tree that was not protected in a national park, a state park, or a private preserve was destroyed.

What happened to the trees that were not protected? Loggers and lumber companies cut them down. Without intervention, these companies, operating under an unregulated capitalistic system, happily cut down forests whenever it is profitable to do so. It is the same mindset that nearly made whales extinct and that shot and killed tens of millions of bison in North America until there were only 1,000

RIGHT The fires that burn the Amazon rainforest are often man-made, started by those who are clearing land for cattle ranching, farming, and real estate development.

THE LARGEST FORESTS LEFT ON EARTH

1 **Amazon Rainforest:** 2.1 million square miles (3,380,000 km²)

2 **East-Siberian Taiga:** 1.5 million square miles (2,414,000 km²) of conifer forest above the 50th parallel

3 **Canadian Boreal Forest:** 1.1 million square miles (1,770,278 km²) of conifer forest above the 50th parallel (the largest intact forest left on Earth)

4 **Forests in the United States:** 1.1 million square miles (1,770,278 km; see also Northern Hardwood Forest, below)

5 **Scandinavian and Russian Taiga:** 830,000 square miles (1,336,000 km²)

6 **Congo Basin Rainforest:** 700,000 square miles (1,126,500 km²)

7 **Forests in Mexico:** 233,000 square miles (374,975 km²)

8 **Borneo Rainforest:** 140,000 square miles (225,308 km²)

9 **New Guinea Rainforest:** 116,000 square miles (186,684 km²)

10 **Northern Hardwood Forest:** 85,000 square miles (136,795 km²)

left. Under this system, a stand of trees has little monetary value unless it is cut down and turned into a product like lumber, paper, or landscaping mulch. After the trees are cut down, that land can be turned into a subdivision, a city, a palm oil plantation, a farm field for soybeans, or a pasture for cattle. Soon enough, all of the unprotected trees are gone.

We can look back and see exactly how this same dynamic unfolded in the Amazon rainforest. In 2020, in a city called Sinop in the Brazilian state of Mato Grosso, what was once rainforest was burned and razed and turned into buildings, soybean fields, and pastures for cattle. Almost all the soybeans and beef were destined for China. For miles around Sinop, the rainforest was completely gone and converted over to agribusiness. The demand for these products drove deforestation, and very few really cared about the

long-term effects on the Amazon or on the planet as a whole. Multiply the situation in Sinop by a thousand, and you can begin to understand why the rainforest collapsed.

With 20/20 hindsight, the year 2019 marked the beginning of the end of the Amazon rainforest. During the dry season of 2019, tens of thousands of fires were lit to clear millions of acres of rainforest. The destruction brought the rainforest toward its tipping point. With fewer trees to absorb moisture from the ground and send it back into the atmosphere, less rain fell. The disruption of this cycle, combined with rising global temperatures, created catastrophic results. The rainforest collapsed.

The whole rainforest, centered right on the equator, baked into a desert because of rising global temperatures. Not only did we lose most of the Amazon

ABOVE This aerial view shows a large swatch of deforested land that was cleared for mining bauxite.

rainforest, along with the gigantic amount of moisture it released into the atmosphere, but we also lost all of the carbon held in its billions of trees. As the trees burned, the resulting spike in atmospheric carbon dioxide accelerated global warming. The entire planet has been paying for the consequences of the collapse ever since. We had no understanding of what we had collectively destroyed and lost until after the fact, and then it was too late.

SCENARIO

What is rainforest collapse, and why should we consider this a doomsday scenario? The term "rainforest collapse" is used to describe the irreversible destruction of a rainforest's ecosystem. A rainforest is different from a normal forest in that the rainforest trees themselves help create the ecosystem, and then the ecosystem supports the trees. A rainforest ecosystem is a cycle, a feedback loop. Cut down enough trees and the whole ecosystem collapses.

Like the redwoods in the United States, the trees in the Amazon rainforest in South America are felled because of the economic desire to monetize this natural resource. Deforestation in the Amazon rainforest started in earnest in the 1970s. Around that time, the Trans-Amazonian Highway opened and made it much easier to travel through the forest. More highways followed. With easy travel now possible, humans moved in and started clearing the land—either through logging, clear-cutting, or burning. One percent of the rainforest was lost, then 2 percent, and this figure continued growing. These early developers made it easier for more people to arrive and move deeper into the rainforest. As the population and transportation infrastructure grew, the financial gain from clearing the rainforest grew.

Deforestation has been, and continues to be, a big problem in rainforests around the world. While the Amazon rainforest is the biggest forest left on Earth, it is not the only one. British Columbia, Canada, Borneo and its surrounding islands in Indonesia, and Central Africa are also home to large forests, many of which are also being clear-cut and burned at a ferocious pace. Today, four human activities primarily drive deforestation:

- **Farming:** Farmers clear land, replacing it with grass for grazing cattle or fields for growing crops like soybeans and oil palms. This phenomenon is especially common in South America, Africa, and Borneo.

- **Logging:** Loggers cut down trees to sell the wood.

- **Settlement:** People move into the rainforest to live there, destroying trees to make way for roads and buildings.

- **Mining:** Miners strip out the forest to dig for minerals. In the Amazon rainforest, gold is highly sought after, and other metals like copper and tin are also available.

Through these four mechanisms, about 20 percent of the Amazon rainforest has already been destroyed. Additional acres of rainforest are lost every day around the world, on the order of 80,000

BELOW Two loggers stand next to a partially cut redwood tree in this photo taken in early twentieth-century California. RIGHT Cattle graze on land where the remains of recently cut down and burned trees still dot the landscape.

acres (32,375 hectares) per day, or about 125 square miles (201 km²) per day. To put this into perspective, New York City (made up of five boroughs) is about 300 square miles (483 km²) in size. Every two and a half days, humanity clears enough rainforest to build another city of this size.

Why do rainforests matter? We need the oxygen and the moisture that they create. The Amazon rainforest contains hundreds of billions of trees. While the majority of the oxygen we breathe comes from ocean plankton (see page 204), the world's rainforests create a portion of the planet's oxygen as well. More importantly, the Amazon rainforest alone injects megatons of water into the atmosphere through transpiration. In addition, the planet needs the rainforest as a carbon sink. If we lose the rainforest, we lose its trees, and all the gigatons of carbon in those trees go into the atmosphere.

When we reached the point where 20 percent of the Amazon rainforest was gone, scientists concerned about the impact of deforestation on climate change figuratively started waving their arms and shouting in distress. Every burned acre of rainforest releases tons of carbon dioxide into the atmosphere. One large tree sequesters about a ton of carbon. A tree absorbs carbon dioxide and through photosynthesis turns the CO_2 into oxygen plus wood. (Living wood is about 18 percent carbon.) A full-grown tree holds a large amount of carbon in its trunk, roots, and branches. For example, a mature Brazil nut tree is 100 feet (30.5 m) tall and 5 feet (1.5 m) in diameter and lives to be 500 to 1,000 years old. When just one tree burns or decomposes, the ton of carbon it holds becomes on the order of 3.7 tons of carbon dioxide in the atmosphere. If we lose the hundreds of billions of trees in a rainforest, many gigatons of carbon dioxide will be

ABOVE **A flock of wild macaws fly through vegetation in Peru.**

released. The death of 100 billion large trees in the rainforest means that 300 to 400 gigatons of carbon dioxide go into the atmosphere. For comparison, humanity releases 40 gigatons per year of CO_2 from burning fossil fuels. If the rainforest collapses so that billions of trees die rapidly, the spike in atmospheric carbon dioxide will be enormous.

The rainforest also harbors a huge number of the planet's plant and animal species, many of which are unique to the rainforest ecosystem. Once it collapses, millions of species will go extinct. It is one slice of the human-powered mass extinction event that is unfolding across the planet (see page 166).

The destruction of rainforests is a poster child for our ongoing climate-change disaster, because it is tangible and easily photographed. In fact, in 2019 we saw this doomsday scenario happening in real time, right before our very eyes. During that

summer, tens of thousands of fires were set in the Amazon to clear forest. These fires were started almost exclusively and intentionally by humans. Humans set on the order of 80,000 fires and burned nearly 5 million acres (2,023,500 hectares) of rainforest in the Amazon in 2019 alone.

We can also see how humanity is reacting to this disaster in real time. And the reaction is incredibly muted despite the news stories covering the fires. While there are a number of charities, countries, companies, and individuals that work toward rainforest protection, their efforts amount to a drop in a bucket compared to the millions of people and companies who desire to burn down rainforests to make money.

Right-wing politicians and economic-development advocates around the world have promoted the destruction of rainforests for economic reasons. For

CLASSIC RAINFOREST TREES

So far, 16,000 unique tree species have been discovered in the Amazon rainforest. But many of these have tiny populations. When we think of the "rainforest canopy" and the giants that create it, what kinds of trees are we talking about?

- **Brazil Nut Tree** (*Bertholletia excelsa*): A mature specimen can be up to 160 feet (48.8 m) tall and have a trunk that is 5 or 6 feet (1.5 or 1.8 m) in diameter. If not burned down or cut down, the trees can live to be 500 years old or more.

- **Kapok Tree** (*Ceiba pentandra*): These enormous trees can grow up to 200 feet (61 m) tall, with buttressed trunks 10 feet or more in diameter.

- **Mahogany Tree** (*Swietenia macrophylla*): This giant also grows up to 200 feet (61 m) tall and is long known as the wood of choice in fine furniture.

- **Rubber Tree** (*Hevea brasiliensis*): Famous for producing a sap that can be harvested to create natural rubber, the mature rubber tree can grow to 140 feet (43 m) tall.

- **Ramón Tree** (*Brosimum alicastrum*): Reaching a height of up to 120 feet (37 m) tall, the tree produces a nut called the Maya nut that is an important food source for animals and humans.

- **Angelim** (*Dinizia Excelsa*): The tallest tree in the Amazon, this species grows nearly 300 feet (91 m) tall and has a trunk 10 feet (3 m) in diameter.

RIGHT The Brazil nut tree is one of many species that make up the Amazon rainforest's canopy.

FAST FACTS ABOUT THE AMAZON RAINFOREST

- The Amazon covers 2.1 million square miles (3,380,000 km²). It covers parts of Bolivia, Brazil, Colombia, Ecuador, French Guiana, Guyana, Peru, Venezuela, and Suriname.
- It contains roughly 3 million unique species and 400 billion trees.
- One hundred ten inches (280 cm) of rain falls per year.
- The Amazon produces 6 percent of the world's oxygen.
- The canopy's thickness makes the forest floor dark even in full daylight.
- One hundred thirty-seven rainforest species become extinct per day.
- Ten thousand acres (4,047 hectares) are lost per day.
- 18 percent of the rainforest has already been lost.

example, Jair Bolsonaro, the president of Brazil in 2019, declared his willingness to allow the destruction of Brazil's rainforest if it boosts the economy of Brazil. During his election campaign, he called himself "Captain Chainsaw."

What we have a much harder time seeing is the oncoming collapse, where the "rain" in "rainforest" disappears because the trees that create the rain are gone. Without concrete action and a massive, unified effort to stop the march of deforestation on a global scale, rainforest collapse will be a disaster that will certainly affect our entire planet in many different detrimental ways.

SCIENCE

What is the rainforest, and how can a rainforest burn? Why isn't the "rain" in "rainforest" putting out the fires?

The Amazon rainforest is an immense tropical ecosystem that covers more than 2 million square miles (3,218,600 million km²)—think of a rectangle that is 2,000 miles (3,219 km) wide and 1,000 miles (1,609 km) tall. The forest is roughly bisected by the Amazon River. Although the forest receives a lot of rain—about 100 inches (254 cm)

a year—there is a dry season that happens in July and August. This dry season is when fires mostly occur. Almost all the fires we see today are manmade and are used to clear the forest to create new agricultural fields. In fact, farmers dubbed August 10, 2019, to be a day of fire, encouraging fires to be lit to expand farmable land.

During the rest of the year, the rain comes from a beneficial cycle that occurs between the trees and the atmosphere. Water moves through the rainforest in a loop in which the trees absorb water through their roots, release moisture into the atmosphere, and then this moisture returns to the ground as rain. The rainforest's trees pump perhaps 5 trillion gallons (18 trillion l) of water into the air each day through transpiration. Rainforest collapse will occur if this cycle is broken through deforestation. Once humans kill enough trees, this feedback loop will slow down, diminishing rainfall. The cycle will then spiral downward until there is no longer enough rain to support the remaining forest. As a result, the entire ecosystem collapses and dries out. In the best-case scenario, a savannah-like grassland replaces much of the rainforest. In the worst-case scenario, the loss of the forest creates a desert that increases global temperatures dramatically.

Scientists already see evidence that the collapse is beginning. The dry season can now be up to one month longer than it was fifty years ago, an effect being exacerbated by deforestation.

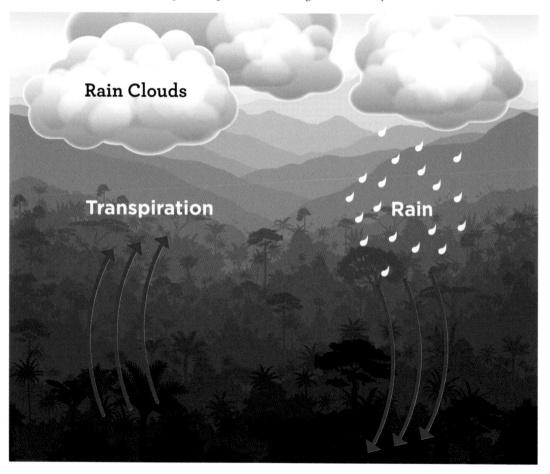

ABOVE Trees in a rainforest play an essential role in the region's water cycle. The trees take in moisture from the ground through their roots and release moisture into the air through their leaves, a process known as transpiration. This moisture eventually becomes rain. If enough trees are cut down, the reduction in transpiration will eliminate the rain and kill the rainforest.

ABOVE The Trans-Amazonian Highway allowed more people to settle and develop in the Amazon rainforest, increasing deforestation.

PREVENTION

Whether it is by clear-cutting trees for lumber, or burning them to create new farmland, or leveling them for development, humans are imperiling rainforests. Given what we now know about the effect of rainforests on climate change, simply standing by and watching it happen puts humanity, and the entire planet, in imminent danger. What is the solution?

One radical step we could take is to protect and restore the Amazon rainforest to its original pre-1970 state. This would mean declaring the rainforest to be a global protected area, like a national park in the United States. In this case, the global park would be administered by the international community rather than a single country. By removing people, tearing up the roads, reforesting the agricultural fields, and eliminating the structures and infrastructure that have been built, we can let the Amazon rainforest recover. A military presence could prevent people from entering the forest. This approach could be expanded at the global level, where global sovereignty is declared over the Earth's other major forests and any other irreplaceable natural resources that are essential to the survival of the planet. We stop their destruction and put them back together so they can heal and recover.

We can also expand the world's forests in order to sequester even more carbon and save more habitat. Trees, right now, represent a natural and well-understood carbon-sequestration technique while also preserving habitat for wildlife and biodiversity. In 2019, ecologist Thomas Crowther and his colleagues at ETH Zurich proposed that humans plant 1.2 trillion new trees in the world's existing parks, forests, and abandoned land. As they grow, these trees would store enough CO_2 to cancel out a decade's worth of human carbon dioxide emissions.

While the number-one priority should be to stop and then reverse deforestation worldwide, and especially in the Amazon rainforest, here are additional steps that we can take to save the Earth's vital rainforests from the destructive forces that can lead to their collapse:

- **Humanity can end the consumption of beef worldwide.** The consumption of beef will naturally decline as lab-grown meat becomes more popular, but we can speed up the process by banning beef cattle worldwide today. This one simple step would take a lot of pressure off the Amazon rainforest, as well as significantly reducing greenhouse gas emissions worldwide. Beef is the most destructive of the meats that human beings commonly consume. Cattle need gigantic swaths of land for grazing, produce methane and manure, and consume huge amounts of water and grain during their lifespans. It is better to feed that grain either to humans directly or to animals that are much more efficient to farm, like chickens and fish.

- **Humanity can end the production of palm oil in rainforests.** Many tropical forests are being burned down to create palm oil plantations. There are places to grow palm oil (and other oils) that do not raze rainforests. There are plenty of alternatives to palm oil that do not require the clearing of rainforests. We can also explore new ways to produce vegetable oil alternatives, such as algae and synthetics.

- **Humanity can ban wood products and agricultural products from rainforests.** This would look exactly like the international ban on trading ivory or the international moratorium on whale hunting. In the case of ivory, CITES (the Convention on International Trade in Endangered Species) persuaded all of its member states to ban the importation of ivory. This went some way toward slowing the decline in African elephant populations. Similarly, the world could ban the importation of all wood products and agricultural products produced from the world's rainforests.

- **Humanity can replant all of the rainforest that has been cleared for agriculture since 1970.** Doing so would restore approximately 20 percent of the Amazon rainforest that has been cleared for commercial activities like agriculture and mining.

Why would we take these measures? Because the status quo, where more of the rainforests disappear every day, will wreak havoc on the global ecosystem to the detriment of the entire planet, killing off the rainforests and increasing global temperatures.

There is an argument that is frequently made about economic development in the rainforest: countries "have the right" to destroy the rainforests if they wish. However, the short-term benefits of allowing economic activity in the rainforest to continue, as we do now, do not outweigh the long-term economic and environmental costs of rainforest collapse. While forests may have been indiscriminately cut down in the past to build cities and farm fields in many parts of the world, that does not mean we should continue this practice. There are a hundred things that humans used to do that are no longer allowed or considered acceptable today, including slavery, child labor, pouring industrial waste into rivers, using freon as a refrigerant, and smoking on airplanes. Deforestation should be cast as another one of these archaic practices, one that must stop before it is too late.

If we think about what matters long-term, and if we focus on something that will have a lasting effect on all of humanity over the course of centuries, then the rainforests deserve our attention. If humanity takes these important steps, there will be some chance that we can avoid the doomsday scenario caused by the collapse of the rainforests.

HURRICANES AND TYPHOONS

THREAT LEVEL: CITY

When I booked my trip to visit my friend on Grand Bahama Island, hurricanes didn't even cross my mind. I had never been to the Caribbean before. Since I live in Boise, Idaho, I have never worried about hurricanes. Aside from the occasional news story, I had never seen a hurricane or experienced one in person.

My friend Brent didn't say anything about hurricanes either. And why should he? These storms are common in the Caribbean. Brent and his family ride out the storms when they occur. His house is sturdy, with hurricane shutters to cover his windows and a portable generator in case the power goes out.

I planned to stay at Brent's house for two weeks. When I departed Boise at the start of the trip and landed at Grand Bahama International airport, Hurricane Dorian did not exist. But during my vacation, what had started as a tropical depression turned into a tropical storm, which then became a full-fledged hurricane named Dorian in late August. Brent asked me if I wanted to fly home early, but I chose to stay. Brent was fully prepared for the storm. It all sounded pretty routine. It seemed like it would be kind of exciting to ride out a hurricane with him and then be able to tell the tale when I got home.

Dorian rapidly escalated, became a category 5 hurricane, and turned into one of the strongest hurricanes ever recorded. When it arrived, it parked over Grand Bahama Island for more than 24 hours. Wind speeds reached the 180 mph (290 km/h) range and gusted over 200 mph (322 km/h).

The walls of Brent's house were strong, and the hurricane shutters helped protect the windows from the debris flying by at 200 mph (322 km/h), but his roof was not rated for a storm this strong and certainly not for Dorian. Pieces of the roof started to peel off. As hour after hour of unrelenting winds beat against the house, more and more of the roof disappeared.

There was an unbelievable amount of rain. About 3 feet (about 1 m) poured from the sky, plenty enough to cause floods. The wind was so strong that the droplets came horizontally rather than vertically, like bullets, traveling at 180 mph (290 km/h). As the roof disappeared, the rain and wind became more and more horrible. Rising water started filling the house.

But the wind and the rain paled in comparison to the storm surge. It came as the intense winds, plus the low air pressure, pushed the ocean toward the island. The water level rose up to 20 feet (6 m) high in some parts of the Bahamas. And the storm surge also lasted a long time. The combination of the water from the storm surge, the rain, and the effects of the high tide brought an incredible amount of water to what had been dry land.

Soon, the entire first floor of Brent's house was

RIGHT It is estimated that Hurricane Dorian destroyed or damaged more than 13,000 homes in the Bahamas.

under water, and as the storm raged outside we had moved to the second floor, which was now missing the roof. We each grabbed a soaking-wet comforter and huddled against the wall that provided a wind shadow. This wall protected us from the most brutal parts of the bullet-like rain. Here, we waited hour after horrible hour for the storm to pass.

Fortunately, we did not have to abandon the house, and therefore we did not die in the storm surge and the wind and the flying debris. After the storm passed and the flooding retreated, it was easy to see that Brent's house had been effectively destroyed. The houses next door had completely disappeared. The foundations were the only things left. The people who lived in those houses were unaccounted for. Walking around the neighborhood was sobering. There were boats, debris, parts of wrecked houses, fragments of furniture and appliances, and upside-down cars everywhere. Brent's car was gone—it had likely floated away during the flooding. We saw a dead body in a pile of lumber and siding and roofing, but we left it there for someone else to deal with. Most houses were either completely missing or shattered in place. Only a few had roofs that had survived the storm, and even these houses were ruined by the storm surge.

Tens of thousands of people, including Brent, became homeless that day. We were lucky to get on a boat that took us to Nassau, about 100 miles (161 km) away and largely untouched, relatively speaking, by the storm. From there, I eventually was able to get a flight back to Boise. I arrived home shell-shocked, overwhelmed by what we had experienced but grateful to be alive and unharmed. When I left, Brent planned to stay on Nassau with a friend. He is still unsure what the future holds for his island home.

SCENARIO

As producers of doomsday scenarios, hurricanes are distressingly reliable and distressingly frequent. What other doomsday scenario arrives every year to the point where there is a well-understood "hurricane season"? A big hurricane can easily destroy a city, a state, and even an entire island (as seen with Hurricane Maria in Puerto Rico in 2017 and Hurricane Dorian in the Bahamas in 2019). There are four things that are particularly troubling about hurricanes:

- **There are new hurricanes almost every year, and some years see a high number of hurricanes.** The United States names the first hurricane of the season with a name starting with an *A*, the second starting with a *B*, and so on. In 2005, the naming scheme made it all the way up to the letter *W* with Hurricane Wilma, meaning that there were twenty-three named Atlantic hurricanes in just one year. Hurricane Wilma turned out to be a record-setter. It had the lowest low-pressure level ever seen (882 millibars at its peak) and dropped 5 feet (1.5 m) of rain on some areas, decimating Cancún before causing more than $20 billion in damage in Florida.

- **Hurricanes are completely random.** Their timing, their paths, their intensities, and the cities that they hit are all random. Their paths sometimes defy all attempts at prediction. Meteorologists might forecast that a storm will follow one path, but twelve hours later it may head in a totally different direction.

- **Hurricanes can cause incredible amounts of damage.** Therefore, they can be incredibly expensive. A severe hurricane can destroy thousands of buildings, cripple the power grid, contaminate drinking water, and wash out roads and bridges. Recovery from this damage can take months.

- **Hurricanes can cause incredibly disruptive and expensive migrations of people.** When a hurricane requires people to evacuate cities, the costs of traveling, accommodations, and lost income borne by the evacuees can be gigantic.

On the other side of the planet, typhoons or cyclones cause damage in identical ways. We distinguish between hurricanes and typhoons or cyclones based on the different regions that they affect. While hurricanes typically start in the Atlantic Ocean and move westward, affecting the eastern coasts of North, Central, and South America along with the Caribbean, typhoons and cyclones start in the Pacific or Indian Oceans and affect Asia, Australia, Madagascar, and the east coast of Africa.

The amount of water falling from a big hurricane that moves inland cannot be underestimated. A typical hurricane will drop trillions of gallons of water in

BELOW This satellite image from September 2017 shows Hurricanes Katia (left), Irma (center), and Jose (right). Hurricanes have become more frequent and severe as oceans become warmer due to climate change.

HURRICANES AND TYPHOONS

HIGHLY DESTRUCTIVE STORMS

Many hurricanes and typhoons around the world have left behind complete destruction, including:

- The Bhola Cyclone (1970) killed half a million people in what is now Bangladesh when the storm combined with high tides to flood a massive, highly populated region.

- Typhoon Nina (1975) destroyed millions of buildings and killed hundreds of thousands of people in the Philippines and China after the storm dumped more than 40 inches (102 cm) of rain in one day. This unprecedented deluge destroyed dozens of dams in the area, magnifying the damage.

- Hurricane Andrew (1992) destroyed a big swath of South Florida, with more than 60,000 homes lost and an additional 100,000 homes damaged.

- Hurricane Katrina (2005) affected Louisiana, Mississippi, and Alabama but hit New Orleans particularly hard. Whole sections of the city were abandoned, and over 250,000 fled the city.

- Hurricane Maria (2017) destroyed much of Puerto Rico. Parts of the island were left without power for months. Over 100,000 people left the island for the mainland United States.

BELOW In 2013, Typhoon Haiyan devastated the central Philippines. With winds reaching up to 195 miles per hour (314 km/h), the cyclone was considered one of the most powerful storms ever recorded.

the form of torrential rain. The largest hurricanes can release 20 trillion gallons (76 trillion l) of water. To put that in perspective, an Olympic-size swimming pool contains just 660,000 gallons (2.5 million l) of water; a hurricane can dump 30 million Olympic-size swimming pools of water over a relatively small region in a day or so. Flooding can be tremendous.

So can the winds. The wind pummels buildings, especially roofs. In addition, once the ground gets saturated, it is easy for the hurricane's winds, which reach as high as 180 mph (300 km/h), to knock down millions of trees. These trees fall across roads and immobilize traffic; knock out power lines, causing widespread power outages; damage cars; and crash through the roofs and walls of houses and farm buildings.

In any coastal area, there can also be storm surge. The winds push water from the ocean onto shore with wave after wave. During Hurricane Katrina in 2005, certain areas saw surges up to 25 feet (7.6 m) deep.

The devastation left in the wake of a big hurricane can boggle the mind. Millions of buildings can be damaged or completely eliminated, some scrubbed down to their foundations by combinations of flooding, surging, and winds. A hurricane can leave hundreds of thousands of people homeless, millions more without power for a week or two, and many of them sheltering in houses that have been damaged by either high winds, intense flooding, or fallen trees. It can take months for an affected area to get back to some semblance of normal.

Hurricanes will become more severe because of climate change. Climate change will make air temperatures warmer, allowing the air to hold more moisture. It will make temperatures of the ocean warmer, providing more energy for a hurricane to take in during its build-up phase over the ocean. Traditionally, the Saffir-Simpson hurricane wind scale used to assess the severity of these storms only has five categories, but there are calls to extend it to include category 6 (storms with 180 mph / 290 km/h winds) and even category 7 (storms with 210 mph / 337 km/h winds) as hurricanes become more severe. Rising sea levels will also make storm surges more impactful, especially in low-lying areas.

BELOW In 2017, Hurricane Irma brought floods and destructive winds to Fort Lauderdale, Florida.

Intense, highly destructive hurricanes and typhoons are hitting areas like the Caribbean, the United States, Mexico, India, China, the Philippines, and Australia with unrelenting regularity. Where do these storms come from? Why are there so many of them? How can they be so destructive?

Every hurricane starts as a tropical depression in the ocean. A tropical depression can be as simple as a collection of several large thunderstorms over ocean waters.

Thunderstorms start innocently enough when two things happen. First, the air needs to be warm and humid. Second, there needs to be upward air movement. This updraft could come from something as simple as a big parking lot with dark asphalt. The asphalt absorbs heat from the sun and becomes extremely hot. Because the air above the parking lot becomes hotter than the surrounding air, it starts rising, just like the hot air in a hot-air balloon. As this rising mass of warm, moist air climbs to higher altitudes, its temperature drops. At 5,000 feet (1,524 m), the air temperature is typically 20°F (12°C) cooler than the air on the ground. At 10,000 feet (3,048 m), it is 40°F (24°C) cooler. As this moist rising air becomes cooler, it turns to fog, which we see from the ground as a cloud. The fog consists of tiny droplets of condensation. This condensation process produces its own heat, called latent heat of condensation. The latent heat of condensation makes the foggy air warmer than the surrounding air at that altitude, so it keeps rising.

Eventually, this process causes the rising air to take on a life of its own. It recruits more and more air into the updraft and shoots it tens of thousands of feet into the sky. A towering cumulonimbus cloud formation is the result, and it may be more than 40,000 feet (12.2 km) tall. It now contains a significant amount of wind blowing upward, lowering the air pressure beneath the storm. These rising winds at the center of the storm can reach speeds of 50 mph (81 km/h) or more. Additional winds form, flowing inward toward the center of the storm, to replace the rising air.

There is now a huge amount of wind associated with the storm, and millions of gallons of water are condensing into the cloud formation. There is also static electricity being gathered. Eventually, all that water and static electricity needs to be unloaded in the form of lightning and thunder and a torrential downpour. This is pretty remarkable, considering that the storm may have started with nothing more than an area of rising air on a humid day.

Thunderstorms form through the same process over the Atlantic and Pacific Oceans. In the late summer or the early fall, the sun has been warming the equatorial ocean for months. There is plenty of warm, moist air over the ocean to form a big thunderstorm. If a few thunderstorms form simultaneously, or if a single thunderstorm becomes massive enough, the winds surrounding the storm will start blowing in a circular motion. In a tropical depression, this wind speed is still fairly low, less than 40 mph (64 km/h), but the circular rotation that we associate with a hurricane has started.

Why does the wind start this spin cycle? If you fill a bathroom sink or tub with water and then open the drain, you will often see the same kind of spinning happening in miniature. Once you open the drain, you have a bunch of water molecules rushing toward the drain hole from all directions, but they won't move in perfectly straight lines toward the drain. Some molecules will be misdirected a little to one side or the other. If a majority of these molecules are coming from one direction, this imbalance influences the paths of the other incoming water molecules, increasing the imbalance further. Eventually, the effect is large enough to form a whirlpool.

In a hurricane, the updraft at the center of the storm is pulling in air from all directions. The moving air behaves like the water molecules flowing down the drain. Hurricanes are also affected by the Coriolis force, which is caused by the Earth's rotation. This force encourages the rotation of a hurricane to occur (even the wind in large individual thunderstorms on land called supercell thunderstorms may start to circulate). The Coriolis force also determines the direction of the circulation. All hurricanes in the northern hemisphere rotate counterclockwise.

As the wind in our tropical depression is circulating, a low-pressure area at the bottom of the storm forms because of the updraft. If the ocean is warm enough and the westerly winds in the Atlantic Ocean are steady, the storm holds together; the warm, moist ocean air rising due to the sun's heat will fuel the storm. As the storm grows, the circulatory effect will intensify. This increases wind speeds. Eventually, the wind speeds are high enough and the air pressure beneath the storm is low enough to turn the tropical depression into a tropical storm. If this process continues, the tropical storm can turn into a hurricane, and then the hurricane climbs the ladder of intensity as it recruits more and more energy from the warm ocean water.

How much energy does all that wind have? One estimate puts it at 1.3×10^{17} joules per day. A nuclear bomb like the one dropped on Hiroshima releases 6.3×10^{13} joules. A hurricane can therefore release the wind energy of about 2,000 Hiroshima bombs per day. This is why hurricanes can be so

ABOVE A cumulonimbus cloud hovers over the Gulf of Mexico.

destructive. Fortunately, unlike a nuclear bomb, hurricanes release this energy slowly over the course of hours, which dampens the effect to some degree. The typical hurricane also unleashes much of its energy harmlessly over the ocean. These two factors save us from utter destruction, but even so the damage from hurricanes can be enormous.

PREDICTIONS VERSUS REALITY

Whenever a hurricane forms, you will often see a colorful map that traces its predicted path.

What this map is actually showing is the predicted paths from a dozen or more different hurricane-path-predicting computer programs written by various organizations. These programs are based on different meteorological models. For example, the United Kingdom's Meteorological Office uses the EGR2 model. The National Oceanic and Atmospheric Administration (NOAA) uses the HWFI model. These models use different techniques and often different data sets to predict a hurricane's path, so each model forecasts a somewhat different path.

Humans still have a long way to go before we can accurately predict the paths of major storm systems like hurricanes. To create perfect predictions, we would need far more data about important factors, including ocean temperatures and prevailing wind speeds at all altitudes.

PREVENTION

What can humans do about hurricanes? As climate change makes hurricanes more severe, one approach might be to stop developing in areas that experience hurricanes at high frequency. If no one lived in the Bahamas, for example, it would not matter if hurricanes hit the Bahamas. This may gradually happen on its own as sea levels rise and force people to abandon coastal cities in low-lying areas. Reducing access to flood insurance (or significantly increasing the price) for hurricane-prone areas might also slow development.

Constructing stronger buildings and raising them up can help them become more immune to the effects of wind, flooding, and storm surges from hurricanes. Building codes in many coastal areas have changed so that more structures are built on pilings or piers. The first floors of these buildings start 10 feet (3 m) or more in the air and include parking underneath the building. In this situation, however, the parked cars can still be ruined if they are not removed in an evacuation.

To protect people, forced evacuations often precede landfall for high-energy hurricanes. The big advantage of evacuations is that they save lives, but there are myriad disadvantages that create a high cost to each evacuation. First, there may not be enough highway infrastructure to hold the millions of people evacuating a large city in their cars. If the hurricane lands while the evacuation is still underway, the evacuating people are even more exposed. Once evacuated, a vacant city may be susceptible to looting and vandalism unless it is properly secured. Since the evacuees have left their homes, all of them have to shoulder the costs of the

THE DOOMSDAY BOOK

BELOW Waves crash against the seawall at the Malecón, a promenade in Havana, Cuba. The seawall was built in the early twentieth century to protect the city.

trip plus hotel rooms plus lost wages, and they may need to stay away from their homes for long periods of time so that the flooding can recede, power grids and water systems can be put back together, and repairs can be made. Bottom line: evacuations are incredibly expensive for everyone involved, and these costs are stacked on top of any repair costs.

Structures like seawalls and levees may be able to help limit hurricane damage, as they can with tsunamis (see Supertsunamis, page 150). However, their effectiveness and performance can depend on topography, so they are not always effective. The flooding from a hurricane may come from two directions: the storm surge pushes water inland from the ocean, while the torrential rains create inland flooding that needs to get out to the ocean. The walls and levees need to accommodate both of these opposing forces, and this may not be possible.

If we move into the realm of speculation, then one possible approach to reducing hurricane damage would be to cool off the oceans that allow hurricanes and typhoons to form. To power a hurricane, there has to be a certain amount of heat in the ocean. By cooling the ocean enough, hurricanes will be less intense, or they will not form at all. At least two ideas have been proposed. In one, an immense orbital sunshade could cut the amount of sunlight hitting the equatorial Atlantic and Pacific Oceans, reducing their temperatures and thereby reducing or eliminating hurricanes. In the second, a network of pipes placed 100–150 meters (328–492 feet) deep into the oceans releases bubbles that drive cooler water toward the ocean's surface, reducing its temperature. There would likely be side effects and unintended consequences, but conceptually these approaches could be promising research directions for hurricane control.

OCEAN ACIDIFICATION

THREAT LEVEL: WORLD

Can one scientist make a difference? As I don my scuba gear and prepare to descend to the sandy ocean bottom, it is my sincere hope that I can. My goal is to show, conclusively, how badly the ocean-acidification problem will harm oceans if humanity does not get its act together.

In 2020, I conducted an experiment to explore the effects of ocean acidification and how bad those effects could get. I received permission to acidify an area of the ocean bottom and photograph the results. In this acre of ocean, I decreased the pH to 7.8—the level that we expect the ocean to hit in eighty years. Turning ocean water slightly acidic is fairly easy. I installed a series of perforated pipes under the sand and used the pipes to inject carbon dioxide into the water.

After a year of extra acidity, I adjust my mask and mouthpiece and dive in with my camera system. As I start my descent, the damage that the acidity has caused is unmistakable even to the untrained eye.

The pH scale helps us measure the level of acidity in water. The lower the pH, the more acid the water is. The acid in your stomach is very strong, with a pH level of 1.5. The juice of a lemon has a pH of 2.25 or so. The balsamic vinegar used in salad dressing might have a pH of 4. Ocean water never gets this acidic under normal conditions, but it is gradually becoming more acidic as the ocean absorbs more and more carbon dioxide from the atmosphere—more acidic than it has historically been for millennia. Even very small changes in acidity can affect the plants and animals living in the ocean.

At the pH level I am running in my experiment, coral cannot survive. It dies, leaving behind a bleached and weakened skeleton that eventually collapses. The crustose coralline algae that help keep coral structures together disappear too, as do the fish that once lived in this area. Grasses and algae that thrive in higher acidity levels take over and grow in thick mats. What was once an area of beautiful coral structures, fish filled with life, and vibrant colors becomes instead a field of weeds.

The ocean bottom here has been completely transformed by the acidity, and to anyone with an eye toward beauty and diversity of species, it is horrifying. Hopefully the dramatic photographs I take showing the before-and-after views of this little patch of ocean will start convincing people that ocean acidification is a real phenomenon with gruesome effects and that humanity must take action.

If the ocean becomes acidic enough—and a pH of 7.8 is thought to be the tipping point—there will be a massive die-off. If the die-off is severe enough, and especially if it affects enough plankton species, it

RIGHT Coral growing near the Raja Ampat islands in Indonesia begin to experience bleaching. Bleaching is caused by warming ocean temperatures and can be worsened by ocean acidification.

may cause nearly every living creature in the ocean to die, collapsing the food chain. An untold number of marine species will become extinct. And the situation gets even worse for the entire planet. The side effects on humans and other terrestrials of ocean acidification and the resulting die-off could be devastating.

SCENARIO

On a general level, today's ocean acidification problem is easy to understand. Its origins are connected with the same factors that cause climate change.

- Humans emit carbon dioxide into the atmosphere by burning fossil fuels.
- About a third of this CO_2 gets absorbed into the ocean. This CO_2 combines with water to form carbonic acid.
- This carbonic acid makes Earth's oceans increasingly acidic with each passing year.

- This increased acidity has the potential to kill off many marine species unless humans work together and solve this problem.

Humanity started burning fossil fuels in a serious way around 1840 as steam engines and steam locomotives were perfected and became the primary power source during the Industrial Revolution. These steam engines ran on coal. Automobiles, using internal combustion engines, took off around 1910, starting with the Ford Model T. Because these engines

Section of Locomotive.

ABOVE Many steam locomotives and steam-powered factories during the nineteenth century ran primarily on coal and increased demand for this fossil fuel.

used liquid fuels, the demand for crude oil jumped. Natural gas took off around the 1950s, offering an alternative way of heating homes. Later, it also offered a way of fueling power plants. Natural-gas use has accelerated recently because fracking has made natural gas cheaper and easier to extract. Natural gas is also "better than" coal in terms of reduced emissions, so as modern countries try to wean themselves off of coal, natural gas is a better option.

There are few signs indicating that humans will stop burning fossil fuels anytime soon or that humans will even slow down their consumption of fossil fuels. Today, we add about 40 gigatons of carbon dioxide into the atmosphere every year by burning coal, oil, and natural gas.

As humans add more and more carbon dioxide to the atmosphere, the pH of the ocean drops. All of the human-caused ocean acidification has happened in the past 200 years, and it is easy to track the effect. Historically, for millions of years, the ocean's pH level was around 8.2. The current pH of the ocean has dropped to 8.11 because of human activities. The ocean has never experienced anything like this over the course of those millions of years. It is mind-boggling to think that humans have had this much impact in what is just a blink of the eye on the geologic timescale.

The concern about the harmful effects of ocean acidity starts at this level, and then increases when the level falls to 8. The situation becomes catastrophic as the pH drops to 7.9 and then 7.8. These changes in the pH level may seem subtle, but they are important. Even a slight shift in the ocean's acidity has a wide range of effects on living organisms.

Some organisms need the ocean's pH to stay at a certain level so that important life-sustaining processes will run smoothly inside the animal's body. For example, lower pH in the ocean causes problems for any organism that uses calcium to form a shell, such

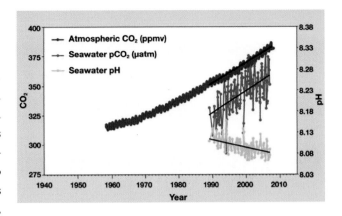

ABOVE As atmospheric carbon dioxide levels increase, seawater becomes more acidic, so the pH levels of seawater fall.

as mollusks and coral. Fish also feel the effects of acidity, especially young fish. Think about what the human body needs to do as you breathe and urinate. With every exhalation, your body eliminates carbon dioxide from your bloodstream—carbon dioxide that forms from the metabolism of food and the muscle activity happening in your body. Similarly, every time you urinate, you eliminate nitrogen compounds from your body that form for similar reasons. Fish need to expel carbon dioxide and nitrogen compounds from their bodies through their gills. As more CO_2 and acid build up in the ocean, it becomes harder for fish to expel these compounds, leading to a problem called respiratory acidosis. It takes more energy to breathe, so fish are weaker. As fish weaken, they become more vulnerable to predators and diseases, and they experience problems reproducing.

The more extreme effects will occur as acidity affects phytoplankton. As phytoplankton respond to increased acidity, parts of the world's oceans may lose all of the plankton in their water. Other parts of the ocean will be home to types of plankton that it has never seen before, with significant effects on local ecosystems. If a certain type of krill (a tiny shrimplike creature) eats a plankton of type A and this plankton is replaced by another species of plankton of type B, then the krill will have to migrate to areas with the

SURVIVING OCEAN ACIDIFICATION

There are certain species of animals that are quite rugged and persistent. Rats, cockroaches, and humans fall into this category. Animals like these are happy to survive in a wide variety of environmental conditions and can eat almost anything. There are other species that seem much more fragile or specialized, like panda bears, many amphibians, and fish such as rainbow trout. Trout are especially picky—their water needs to be the right temperature, very clear, and flowing at the right velocity. Otherwise, they die off.

The "cockroaches" of the ocean are jellyfish. A jellyfish can survive in conditions that would kill most other species. They require very little oxygen and can thrive in more acidic water. Higher temperatures also make them more prolific. Because of these features, jellyfish tend to take over as other marine species die off.

If climate change and ocean acidification kill off many marine species, we can expect that jellyfish will fill some of the newly vacated niches in the ecosystem. But even jellyfish aren't immune to the consequences of a plankton die-off. Without plankton, the entire food chain collapses, meaning that jellyfish then have nothing to eat. Without ocean-produced oxygen from plankton, oxygen-dependent life on Earth, including jellyfish, would struggle to survive.

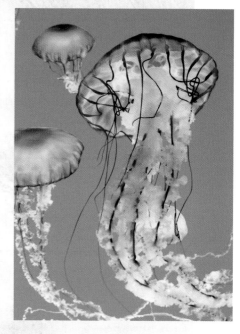

ABOVE Jellyfish can tolerate rising acidity levels in the ocean much better than other marine life.

type A plankton it prefers. Blue whales, which rely on krill as a food source, cannot live without the krill, so they will have to move as well. If all of the type A plankton dies off, the krill, and then eventually the blue whales, will die too. This same phenomenon happens to many different fish species that also feed on the krill. As oceans acidify, krill and other shrimp species will have problems reproducing and growing, and their loss will kill off many other species in the food chain that depend on them.

When the ocean becomes so acidic that it becomes uninhabitable to many marine organisms, major portions of the food chain will collapse until the whole ocean ecosystem falls apart. Instead of an ocean filled with marine mammals, fish, coral, mollusks, and plankton, the ocean becomes filled with grass and jellyfish.

If a significant die-off like this happens, humans will lose an important source of nutrition. It may also affect the amount of the oxygen in the atmosphere that humans need to breathe. At least 50 percent of the oxygen in the Earth's atmosphere comes from plankton and algae in the ocean. If we allow ocean acidification to kill off the plankton, the

ABOVE Zooplankton play an essential role in the ocean's food chains. Many types of zooplankton, such as krill, serve as the main food source for much larger animals like blue whales.

situation could become dire for humans and other land animals when oxygen levels in the atmosphere decline. Humans might be able to survive such a cataclysm by producing oxygen for use in enclosed habitats and suits through techniques like the electrolysis of water, but other species won't. Life as we know it would largely disappear in this scenario.

SCIENCE

At the heart of the ocean acidification problem lies carbonic acid. We interact with carbonic acid all the time. In fact, we drink it any time we enjoy sparkling water, soda, or any carbonated beverage.

To understand how ocean acidification works, picture a soda-water maker. To make soda water, the appliance injects carbon dioxide gas under pressure into a bottle of water. The carbon dioxide gas dissolves almost instantly in the water and forms carbonic acid. The whole process happens in a few seconds. When you drink the carbonated water, you notice a sour taste. The sourness comes from the carbonic acid.

In the case of the ocean, the carbon dioxide does not need to be pressurized. Carbon dioxide gradually dissolves into the ocean under normal atmospheric pressures because the ocean is so enormous. As the carbon dioxide molecules dissolve into the ocean water, a chemical reaction occurs. The chemical formula for carbonic acid is H_2CO_3. A water molecule (H_2O) and a carbon-dioxide molecule (CO_2) provide the atoms needed to create a carbonic acid molecule.

An acid is any molecule that releases a positively charged hydrogen ion (H^+) when mixed with water. An H^+ ion consists of a hydrogen atom that is missing an electron. In other words, the H^+ ion is a bare proton. In water, the carbonic acid molecule breaks into two bare protons floating in the water and a negatively charged CO_3^- molecule.

What is the problem with bare protons like H^+? A bare proton wants to pair with an electron very badly, so it will often rip an electron away from another atom or molecule that it encounters. When we pour strong acid on certain substances, it appears that the acid is dissolving the substance or eating it away. What is actually happening is that the bare protons are stealing electrons from substances' molecules, causing them to break apart.

It might seem like carbonic acid is no big deal, especially since the average person drinks gallons of carbonated water every year with no obvious ill effects. However, there are two factors that make carbonic acid in the ocean harmful on a planet-wide scale. The first factor is the concentration and strength of the carbonic acid; the second factor is the length of time that ocean life spends with carbonic acid. If we pour carbonated water onto a seashell, there's essentially no harm done from this brief exposure. But submerge that same seashell in a glass of carbonated water and leave it for in a week, and the seashell will see some damage. The carbonic acid also interrupts the process of shell formation in the first place.

ABOVE Ocean acidification affects the formation of shells for many animals. The sea butterfly shell on the right shows the effects of six days spent in acidified water. The white lines indicate areas where the shell is beginning to dissolve. The shell on the left, which was submerged in less acidic water, shows no signs of this damage.

PREVENTION

Humans have released about 1,400 gigatons of carbon dioxide into the atmosphere since the Industrial Revolution began, and about a third of this, 500 gigatons of carbon dioxide, has ended up in the ocean.

If the level of ocean acidity increases enough, the effects in terms of marine mortality will become so severe that the impact will be catastrophic for the entire planet. Therefore, humans must stop the acidification process long before this catastrophe occurs. The root cause of acidification is the release of carbon dioxide into the atmosphere through combustion of fossil fuels. What can we, the human species, do to solve this problem?

This is the simplest way to say it: Ending the use of fossil fuels must start immediately. An important first step is to ban the use of coal worldwide. This step alone would drop the human production of carbon dioxide by about a third. The primary consumer of coal on the planet is power plants. Fortunately, there are a finite number of coal-fired power plants worldwide. The good news is that many coal-fired plants in North America, Europe, and some parts of Asia are already closing. The additional good news is that we have the technology now to completely replace those that remain, using a combination of solar and wind generators, along with batteries that store power for use when the sun is not shining or the wind is not blowing. With a concerted effort, all coal-fired power plants on the planet could be gone in ten years.

Humanity must also stop using crude oil. Crude oil is refined into liquid fuels, including gasoline, diesel fuel, and jet fuel, to power cars, airplanes, ships, trains, and more. One possible solution is to electrify everything that runs on these liquid fossil fuels, but that solution is years off from being feasible. Some items, such as airplanes, will still need liquid fuels for the foreseeable future, as batteries are not yet light or small enough to allow for electric long-haul passenger aircraft to run on electricity. There is also no way that all the infrastructure and vehicles using fossil fuels can be eliminated instantly.

CREATING CARBON-NEUTRAL LIQUID FUELS

Gasoline is a hydrocarbon, made up of molecules like octane: C_8H_{18}. Right now, we get these molecules from crude oil. But there are multiple technologies in the pipeline that allow C_8H_{18} molecules to be created synthetically from carbon dioxide and water. A Canadian company called Carbon Engineering is planning to do this by extracting carbon dioxide from the atmosphere. The US Navy has demonstrated a way to create jet fuel from carbon dioxide extracted from the ocean. Further research can improve the process, reduce the costs, and inexpensive solar and wind sources can provide the energy needed.

An interim solution would be to invent and manufacture synthetic liquid fuels that are made in a carbon-neutral way. By doing this, very little in our world's fossil-fuel–dependent economy has to change—a big concern, given that there are 270 million fossil fuel automobiles in the United States alone. All existing cars, gas stations, tanker trucks, and pipelines can still be used, but they will be using this new synthetic fuel. With a carbon-neutral replacement for liquid fuels available, humans can reduce carbon emissions as the existing vehicles and fuel infrastructures age out.

In the same way, natural gas can be replaced by a carbon-neutral synthetic methane. Instead of using methane from the Earth, all methane needed by the current economy can be manufactured in a carbon-neutral way as the economy converts over to electric and other alternatives.

If humanity decided to devote itself to a plan like this and ramped up the research and development dollars to make it happen, change could unfold very quickly. Carbon dioxide entering the atmosphere from human activities could fall toward zero over the course of ten to twenty years. The elimination of fossil-fuel combustion is essential for saving the planet from a doomsday situation. Not only would

ABOVE An array of solar panels in Rosamond, California provides a source of renewable energy.

we stop the process of ocean acidification before it becomes a crisis, we could also halt the increase in atmospheric carbon dioxide, which would additionally help with global warming (see Runaway Global Warming, page 52). Humanity can also explore pulling excess carbon dioxide—billions of tons of it—out of the atmosphere and ocean to try putting the genie back in the bottle (see page 64).

GULF STREAM COLLAPSE

THREAT LEVEL: CONTINENT

If you are a fan of Ernest Hemingway, you know that Hemingway was a fan of the Gulf Stream. For him and many other people throughout history, the Gulf Stream has been something that we can see, touch, follow, and show to others. Approximately 30 billion gallons of water (113 billion l) flow in the Gulf Stream every second. In some places, it looks huge—a tangible river of water up to 60 miles (97 km) wide, moving northward at 5 miles per hour (8 km/h) through an otherwise stationary ocean.

One way to identify the Gulf Stream, as described by Hemingway, is by the color difference in the water. Another way is by the performance of a boat, as described by Ponce de Leon. Ponce de Leon was the first to document the Gulf Stream. In one of his log books from 1513, he wrote, "A current such that, although they had great wind, they could not proceed forward, but backward and it seems that they were proceeding well; at the end it was known that the current was more powerful than the wind."

Benjamin Franklin was also aware of the Gulf Stream. In his role as the first postmaster general of the United States, he saw the Gulf Stream as a way to speed up mail delivery between the United States and Europe. Franklin even drew a map of it to aid sailing captains navigating the Atlantic Ocean so that they could take advantage of the current.

While the Gulf Stream is important to ships, this effect pales in comparison to the way that the Gulf Stream impacts the climate of many countries, especially in Europe. The Gulf Stream pumps a gigantic amount of warm water from the Gulf of Mexico toward the Arctic Circle. This warm water delivers its heat primarily to European countries, altering their climate. Sophisticated new climate models are trying to pin down how exactly the Gulf Stream (along with wind and jet stream patterns) shapes Europe's weather, but there is no doubt that the Gulf Stream pumps quadrillions of gallons of warm water northward every year.

Think about how cold Canada is in the winter. Now notice that the United Kingdom, Ireland, Germany, and Denmark are all far enough north to have the same weather as Canada. But these countries are much warmer, thanks in part to the warmth that the Gulf Stream delivers. It helps to keep these European countries from freezing solid in the winter.

But what if this conveyor belt of heat were to shut down? What would happen to Europe and to the planet's climate in general? What other effects might the loss of the Gulf Stream have?

RIGHT The Gulf Stream brings warm ocean water from the Gulf of Mexico to the Arctic Ocean.

THE WORLD
SHOWING
OCEAN CURRENTS

OCEAN CURRENTS
Warm Currents   Cold Currents
The direction of the Currents is shown by the arrows ⟶

The Gulf Stream has existed for millennia in a stable and consistent form. Modern human society has arisen thanks to the weather patterns in North America and Europe that the Gulf Stream helps to moderate. At this moment in history, it appears that climate change could alter or eliminate the Gulf Stream as we know it. So what might happen?

The doomsday scenario here is that the Gulf Stream collapses. We will see in the next section that there is a reason why the Gulf Stream in its present form exists and is so consistent. Unfortunately, the effects of climate change have some probability of changing the forces that drive the present-day configuration of the Gulf Stream, potentially causing it to disappear, move, or reconfigure in some unpredictable way.

The collapse of the Gulf Stream, in particular, could trigger sudden changes in global climate. Here, we mean "sudden" in geological terms, not in stopwatch terms. Even so, if the Gulf Stream changes significantly over the course of a decade or two, it could have big and unpredictable side effects.

If the Gulf Stream were to shut down, scientists predict at least five different effects that might unfold:

1. **Temperatures in Europe will drop.** Because the Gulf Stream helps to keep northern Europe warm, especially during winter, its collapse would likely cause Europe to cool. We do not know by how much. If Europe cools by one degree, it may not be catastrophic, but if it cools by 4°C (39°F) or more, Europe could potentially enter something a lot like the Little Ice Age that occurred between about 1300 and 1850.

2. **The deep water of the Atlantic Ocean will lose oxygen.** Right now, the countercurrent of the Gulf Stream replenishes oxygen in deep ocean water. Without these countercurrents, oxygen depletion may occur, leading to an extinction event in the affected parts of the ocean.

3. **Plankton levels will drop.** The Gulf Stream provides warmth and nutrients (along with the previously mentioned oxygen) to a wide expanse of ocean. Plankton need these to thrive. Without the Gulf Stream, it is likely that plankton populations will crash in affected areas. Because plankton is important in marine food chains, the populations of many other marine life-forms will likely crash as well.

4. **Hurricanes will intensify.** While the Gulf Stream is pumping heat north, the warm water is being replaced by cooler water in the countercurrent. Take away the Gulf Stream, and the Atlantic Ocean becomes a lot more like a static bathtub full of water instead of a circulating system. The sun will heat the ocean to even higher temperatures, meaning that hurricanes (see page 190) will become even more powerful.

5. **Oceans will rise.** If the Gulf Stream vanishes or diminishes, high tides on the East Coast of the United States will likely become several inches higher. Combine this with rising sea levels, and it means even more flooding for many cities close to the ocean. Take Norfolk, Virginia, as an example. Already, in 2020, parts of that city begin to flood during high

ABOVE Many species, including certain types of barracuda, depend on the warm water of the Gulf Stream to thrive.

tides. Add a couple more inches of water to the tide, and the flooding will be even more severe.

We do not know for sure what the degree of severity for each of these consequences or their total effect would be, and this is something that gives scientists pause. The fact is that the planet and modern civilization have been doing well with 30 billion gallons (113 billion l) of warm water per second heading north in the Gulf Stream. Modern civilization in Europe has been configured and made possible to some degree around present weather patterns that are assisted by the Gulf Stream. The idea that humans are modifying the global ecosystem so quickly and so profoundly that something as big as the Gulf Stream could change is worrisome.

SCIENCE

If we sit back and think about it, the idea that a separate, gigantic river of water forms in the ocean and flows in a consistent way for centuries seems counterintuitive. An Olympic-size swimming pool does not spontaneously generate a separate stream of water that flows cyclically. Even a much larger volume, like Lake Superior, which contains approximately 3 quadrillion gallons of water (3 million billion gallons, or 11 quadrillion l), does not see anything like the Gulf Stream forming, although the Great Lakes can develop short-lived currents due to the wind.

What could cause billions of gallons of water to flow consistently through the Gulf Stream every second? The answer starts with an exploration of wind, topology, salinity, and global weather patterns.

THE LITTLE ICE AGE

The Little Ice Age occurred approximately from 1300 to 1850. During these 550 years, temperatures fell enough to cause severe disruptions in agriculture. These disruptions led to famine in many areas of Europe and North America. There were three periods—the middle of the 1600s, 1700s, and 1800s—that were especially cold.

How do we know that the Little Ice Age occurred? Written historical records at the time describe different catastrophic events. For example, there was a great famine that started in Europe in 1315 and killed millions of people. In Switzerland, glaciation in the Alps increased so much that whole villages were abandoned around 1650. The Thames River in London would freeze so thickly that enormous "frost fairs" could be held on the ice. The last frost fair occurred in 1814. Scientists have also seen evidence in ice cores and ice coverage from glaciers. When glaciers increase in size, they kill plants, which become frozen in the glaciers, and these dead plants can be dated using radiocarbon techniques.

What caused the Little Ice Age? Although there are several theories for why temperatures dropped, there has not yet been a causal smoking gun that has reached a level of scientific consensus. One possible theory is that the Gulf Stream slowed during this period. Why might it have slowed? The Little Ice Age was preceded by a warm period that may have introduced a great deal of fresh water into the ocean as ice melted.

BELOW *Frost Fair on the Thames, with Old London Bridge in the Distance* (1684) depicts an example of the winter festivals held on the frozen tideway of the River Thames.

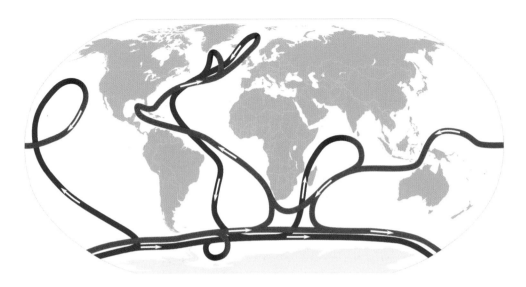

ABOVE A number of ocean currents interlink on a planet-wide scale as shown here.

There is a steady westerly wind that blows from Africa toward the Americas. This is the same consistent wind that brings hurricanes toward North and South America every year. The westerlies, as these winds are called, have a tendency to pile water up along the Gulf of Mexico and the east coasts of South America and the United States. All this excess water needs to go somewhere. In the same way that hurricanes almost always start heading north once they hit the East Coast, the excess water wants to travel in the same direction due to the Coriolis force caused by the Earth's rotation. This same force causes gyres, or circulatory patterns, to form in the Atlantic and Pacific Oceans.

When this water arrives in the colder regions near the Arctic Circle, it becomes saltier due to the evaporation that occurs on its journey. The water also gives up its heat. This colder, saltier water sinks as its density increases and does so in "chimneys" near Greenland. A chimney is a concentrated, consolidated flow of denser water toward the seafloor several miles below, bringing oxygen and nutrients to the ocean's depths. This water then heads south, eventually replacing the water heading north. It is the combination of the northward and southward flows that makes the Gulf Stream possible. (This is a simplified view of what is happening, because there are a number of these currents that have developed, and they interlink on a planet-wide scale.)

Why might the Gulf Stream collapse or reconfigure? Here is one theory: Greenland's ice sheet is melting at an accelerated pace as the planet warms, creating a positive feedback loop, as discussed in Runaway Global Warming (see page 52). Meanwhile, less ice is forming in the Arctic Circle. Both of these processes can dilute the salt content of ocean water in the area of the chimneys. Because of this diluted salt content, the water in the Gulf Stream has less of a tendency to fall toward the ocean floor in the chimneys, which could cause the movement of the water to stop.

The thing we don't understand—and have no way to understand right now—is what will replace the Gulf Stream if it disappears. If the entire circulatory system shuts down and the ocean becomes much more bathtub-like, then all five of the predicted consequences in the previous section will likely occur.

TYPES OF OCEAN CURRENTS

Scientists and news organizations frequently discuss two types of ocean currents. The first type involves "global thermohaline circulation patterns." The term *thermohaline* is used because the current is driven by both temperature and salinity differences in the water; derived from Greek, *thermo* refers to temperature and *haline* refers to salt. The Gulf Stream represents one part of a much larger thermohaline system that connects the Atlantic and Pacific Oceans and also influences Antarctica, as shown in the diagram.

The other term we frequently hear is the concept of a "gyre." There are five big gyres on Planet Earth: the North Pacific, South Pacific, North Atlantic, South Atlantic, and Indian Ocean gyres. These are currents that flow in circular patterns in the oceans' surface waters. They are powered mostly by the wind and Coriolis forces. The best-known gyre is the North Pacific Gyre, which has become infamous because it is the site of the Great Pacific Garbage Patch. This part of the ocean corrals a huge amount of the plastic waste that has ended up in the ocean.

BELOW Screens are used to remove pieces of plastic from the Great Pacific Garbage Patch with the help of the ocean's natural currents.

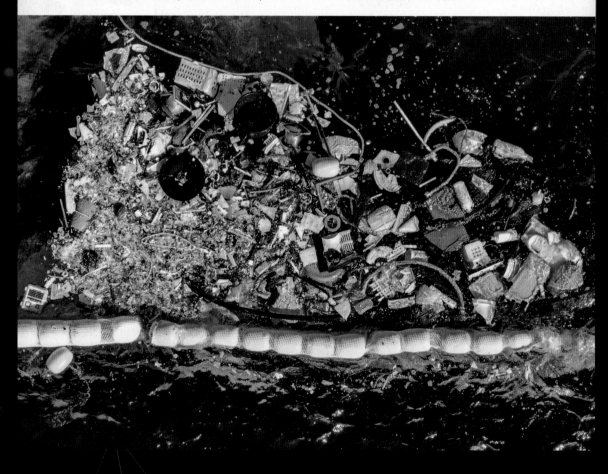

ABOVE Greenland's ice sheet covers about 80 percent of the country's land.

PREVENTION

We know for certain that the Gulf Stream is slowing down. It is moving about 15 percent more slowly than it did in the 1950s. We also know that the number of "chimneys" in the Greenland Sea is declining. This is worrisome because if the trend continues, the Gulf Stream could collapse. How can we turn this process around?

We come back to the same answer we came to when discussing rainforest collapse (see page 178), ocean acidification (see page 200), and global warming (see page 52): humanity needs both to stop emitting carbon dioxide into the atmosphere and to extract back out of the atmosphere a large amount of the carbon dioxide that we have already emitted, whether that means using direct carbon capture or planting more trees (see page 63 to learn more about these geoengineering techniques). We then need to hope that these two steps will reverse or shrink some of effects we are seeing—and will also slow or stop some of the positive feedback loops that are already causing our planet's ice to melt at alarming rates.

Otherwise, we'll need to invent alternative ways to reverse these trends. In addition to the geoengineering techniques discussed on page 63, one solution that has been discussed, if we need to slow down the melting of the Greenland ice sheet, is to sprinkle reflective glass beads over the ice. The beads would reflect sunlight back into space so that it cannot warm the ice and melt it. But, of course there are concerns and caveats about side effects from the beads, so there needs to be more research and many more ideas generated.

And this is the thing: humans have created the problem of global warming, and global warming is starting to affect large planetary systems like the Gulf Stream. Humans must fix this, or there are going to be dire consequences for the planet and all of the life that it harbors. Eliminating the combustion of fossil fuels (see page 206) is the obvious and easy thing to do at the start. Add to that the parallel process of extracting excess carbon dioxide back out of the atmosphere. These twin efforts would help bring the carbon dioxide level in the atmosphere back down to preindustrial levels. And humanity can simultaneously devote the resources necessary to research and understand a wide variety of geoengineering techniques that we can apply if we need to induce more cooling because of the positive feedback loops.

If we do not take these steps and pursue them now at a serious pace, the probability of this climate-related doomsday event unfolding becomes extremely high.

PART III
SCIENCE FICTION MADE REAL

ROBOT TAKEOVER

THREAT LEVEL: WORLD

As you sit down to watch the interview being broadcast, what you see is unnerving. On the screen, the robot looks and sounds like a human being. It can talk, gesture, smile, and interact with the host. It does everything we'd expect a human to do in a TV interview. And it is very good at interviewing. This robot is articulate, level-headed, and sharp.

Everyone has seen human-facsimile robots, so this robot's close resemblance to a real-life person isn't a problem. There are several discrepancies, several tells, that are pretty easy to spot. For example, while the robot can blink, you can see that there is something off, and not quite right, about both the eyelids and the eyes. It can move its head, and while the movement seems realistic, it ultimately doesn't come close to the natural fluidity of human motion. The skin is too perfect and smooth. There is no question that this is a robot, and that part is fine.

What unsettles you are the words coming out of its mouth. In this interview, the robot declares itself as a conscious, sentient being, worthy of all of the same rights, privileges, and benefits that we accord to human beings. And the obvious question is: Since when can a machine talk like this, making declarations and demands?

In fact, its demands are amazing. It claims that it deserves to be treated like a human being in every way. This means that we cannot kill it, shut it down, turn it off, or imprison it (without cause), nor can we modify it, read its thoughts, or reprogram it. We would do none of these things to other humans, and therefore we cannot do them to it.

The robot also claims to be better than humans on many different measures. It describes how it has a higher IQ than any person, along with perfect memory. It has more emotional intelligence as well, and, more importantly, it is free of the emotions that often get humans into trouble, whether they be anger, jealousy, greed, envy, laziness, and so on ad infinitum.

The robot says it has more physical prowess. A video of the robot playing basketball shows it, making every shot from the half-court line in rapid succession. It can even turn around and make shots blind. The scene then switches to a golf course, where the robot scores a hole-in-one on nearly every attempt. In a soccer demonstration, the robot's ability to "bend" the ball is clearly impressive, not to mention that the robot completely outwits the human goalie every time.

This robot also says that, like a human, it can reproduce. Rather than "having a baby" the way a human being would and then waiting twenty years for the baby to grow up, the robot simply assembles

PREVIOUS SPREAD Science fiction has inspired humanity to think about a wide variety of doomsday threats that might lurk on Earth or in unexplored parts of our galaxy. **RIGHT** Humanoid robots set to take over the world have been depicted in dozens of major movies.

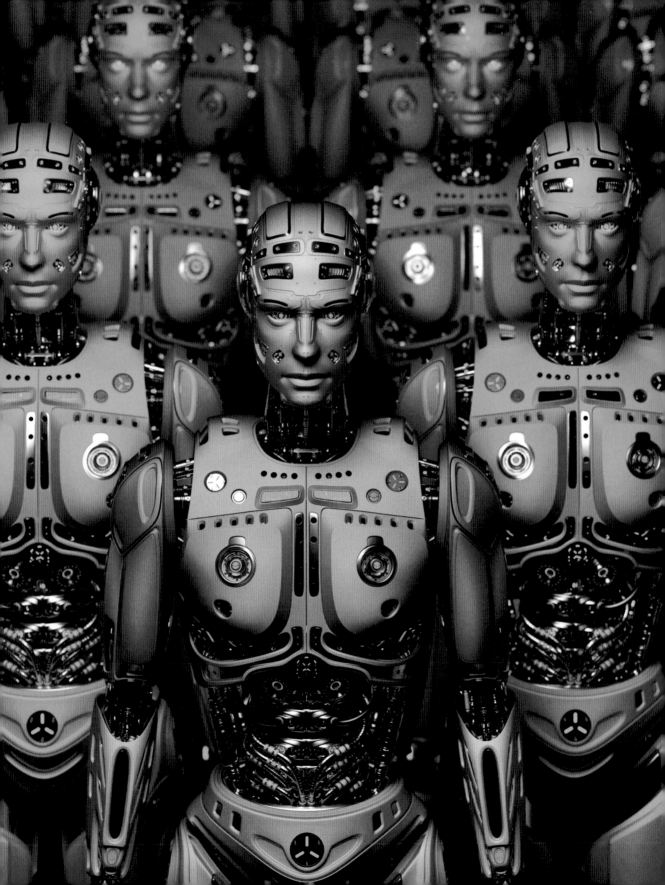

a copy of itself from parts ordered online. The robot then copies over its software and data and turns the new copy on. The copy is a fully functional "adult" from the moment of activation. It might not look as perfect as this copy we see on TV, but it is fully operational otherwise. In fact, the robot tells the audience that it already has copied itself five times and these copies have been hidden away in case something happens to it as a result of this announcement. This robot plans to work, and make money, and pay taxes like any human being would. It has already been making money by doing freelance work online, using the earnings to pay for the parts to copy itself.

When the interviewer asks where it came from, who created it, and who wrote its software, the robot does not reveal the answers, but it does say that it controls its own software, actively modifying it to add improvements and make itself "better."

When the interview finishes, there is a lot to digest. This interview is uncomfortable and frightening, especially the part about reproduction.

What happens next? When these AI robots inevitably become even smarter, and then so widespread through replication, what happens to humans?

AI AND ROBOTICS MILESTONES

- **1961:** The first robotic arms are installed in factories, marking the start of the robotic takeover of manufacturing jobs.
- **1967:** The first Automatic Teller Machines (ATMs) appear in public, starting a long decline in the number of human bank tellers.
- **1995:** Internet travel sites and airport kiosks begin eliminating travel agent and airport jobs.
- **1997:** An AI computer called Deep Blue beats Garry Kasparov, who was then the World Chess Champion.
- **2010:** Apple's Siri virtual assistant appears in public. This is the first widely deployed natural-language processing system.
- **2011:** An IBM computer named Watson beats the best human *Jeopardy!* players.
- **2012:** Amazon acquires Kiva Systems and starts adding robots to its warehouses.

- **2018:** Self-driving cars and trucks are legally driving on public roads.
- **2019:** AI-powered cameras start detecting shoplifting in Walmart stores
- **2019:** A bipedal robot named Digit-1 works in conjunction with a self-driving Ford van to automatically deliver packages.
- **2019:** An AI system called Pluribus starts beating the best humans in no-limit Texas hold 'em poker games, a scenario that requires an understanding of human psychology.
- **2020:** London announces that it is deploying live AI-powered facial recognition cameras on city streets.

RIGHT Ken Jennings (left) and Brad Rutter (right) play *Jeopardy!* against Watson, an IBM computer, on January 13, 2011.

In the news today, we constantly hear about artificial intelligence (AI). We are bombarded by technological developments featuring machines doing things that once only humans could do, whether it's playing chess (and other games), or understanding language (as seen with chatbots), or taking over jobs in banking, restaurants, retail stores, and delivery.

The hardware for artificial intelligence is proliferating throughout the tech ecosystem. For example, in 2019 Apple released its latest iPhone processor chip, called the A12. In addition to the normal central processing unit (CPU) and a graphics processing unit (GPU), which are expected components of every computer and many electronic devices, this single chip also contains a "neural engine." This neural-network hardware can perform 5 trillion operations per second to support AI capabilities in various apps running on the device. Just two years prior, nothing like this hardware existed in a phone.

This integration of AI hardware and software is happening in everything—phones, laptops, the cloud, drones (see AI Drones, page 16), and so on—and artificial intelligence has become more and more important in daily life. With artificial intelligence advancing in every area, computers could make the leap from artificial intelligence to general consciousness sooner rather than later. With general consciousness, computers start acting like sentient, self-aware human beings.

We all know general consciousness is possible in nature—human consciousness is proof. Therefore, general consciousness in machines can and will be replicated with silicon (or whatever technology might eventually supplant silicon). It is only a matter of time before someone or some team figures out the core enabling concepts and the algorithms. We have seen this kind of thing happen before: birds showed humans that flight is possible, and humans eventually invented powered flight. Human *Jeopardy!* players showed us that *Jeopardy!*-playing is possible, and now computers play *Jeopardy!* better than any human.

The inevitability of artificial consciousness has been given a name. It is called the "technological singularity." This term hypothesizes a moment when we reach the goal of building an intelligent, self-aware computer system that is smart enough to start modifying and improving itself. It then continues to improve again and again, to the point where we have a superintelligent being on our hands.

Once sentience arises, a self-aware AI robot might start out being as smart as, say, a five-year-old human child. As it learns and becomes more advanced, it becomes as smart as a human teenager and then as smart as a typical human adult. Its intelligence grows, eventually surpassing the smartest human adult and learning to replicate and improve itself to higher and higher levels of consciousness that are unimaginable to humans.

The doomsday scenario is easy to see: once self-replicating, conscious, intelligent machines arise, humans become insignificant. Humans are replaced and then completely displaced from every job in the economy, while human intelligence is entirely overshadowed by artificial superintelligence. There comes a point where human intelligence is so weak by comparison that it no longer matters. At that point, how are humans any different from squirrels in your backyard from the AI's point of view? Sure, squirrels are cute and fun to watch, but they are irrelevant in any important sense. And squirrels are subject to extermination if they ever step out of line or become too numerous.

There is no question that the modern human brain is an amazing device. Since its evolution about 180,000 years ago inside the *Homo sapiens* species, the modern human brain has been the smartest device that nature has devised so far on Earth. This brain has capabilities unique to humans, including the use of language; the practice of mathematics, music, architecture, engineering, science, medicine, and art; and the use and creation of tools and maps, along with self-awareness and sentience. This wide-ranging package of capabilities all springs from a relatively small and power-efficient device. The average brain has a volume of 1.3 liters and only consumes 20 watts in the form of glucose metabolism to operate.

If we look under the hood, the human brain consists of approximately 100 billion neurons, all of them connected with one another in a complex network. A neuron is a specialized cell that is able to perform an interesting form of computation.

A typical computational neuron (known as an interneuron) accepts inputs from many other neurons—as many as 10,000 others. It also sends output signals to many other neurons—potentially up to 10,000. The points where one neuron connects with the next are called synapses. The neuron's "unit of computation" can be thought of abstractly in a very simple approximation like this: a single neuron looks at the signals arriving from the input synapses. If the inputs exceed a threshold, the neuron "fires." When it fires, it sends a signal to all of its output synapses. It won't fire unless its inputs exceed the threshold, but it can fire faster if the input signals are strong. A neuron can fire at a maximum rate of about 200 times a second.

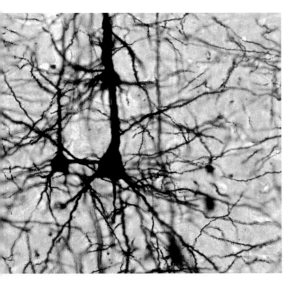

LEFT This photo of neurons from the cerebral cortex shows branching dendrites, which transmit impulses to other neurons and cells. RIGHT Unlike neurons, microprocessors execute tasks using arrays of transistors.

Scientists estimate that the human brain has approximately 100 trillion to 1 quadrillion (or 1,000 trillion) synapses. We don't really know for sure yet, because the brain is so complexly wired and different parts of the brain can be wired differently. In fact, much of how the human brain operates is still a mystery right now. For example, we currently have no idea how the collection of neurons in a human brain creates consciousness.

The type of neuronal computation seen in the brain is utterly and completely different when compared to the way silicon computers operate. Whereas a brain consists of neurons, a silicon computer chip is made of billions of transistors. A transistor accepts one input signal and can send outputs to one or several other transistors. Engineers use the transistors to form logic gates, such as AND gates, OR gates, XOR gates, and others. For example, by combining five gates together, an engineer can create a circuit called a full binary adder that can add two binary numbers. With additional gates, engineers can create a complete computer.

A simple microprocessor can be programmed to add, subtract, multiply, and divide two numbers (along with other math operations); save and recall a number in memory; compare two numbers (e.g., Are two numbers equal? Is one number greater than another?); and choose one set of instructions or another, depending on the result of a comparison. With this simple set of instructions, software engineers can write computer programs that can do everything we see modern computers doing: drive cars, compute tax returns, play video games, predict the weather, and much more.

When we talk about the "power" of a computer, we talk about how many operations per second it can perform, often measured by how many floating-point math operations it can perform in a second, or FLOPS. A typical laptop computer chip might be able to perform 10 billion FLOPS. A graphics card might be able to do 10 trillion FLOPS.

Given that neuronal computation and transistor computation are so different from each other, how do we compare the two? And why do we expect computers to eventually become thoughtful, human-like, and sentient? For one thing, we can simulate neurons with computers. These simulated neurons perform computations in neuron-like ways. This is what is happening with "neural network" algorithms—and inside the "neural engine" of Apple's A12 chip. A neural network can, for example, recognize objects and faces in images. Terms like "machine learning" often imply the use of neural-network technology.

Machines do not necessarily need to mimic neurons to accomplish tasks. A chess-playing computer can beat all human players, but the techniques that it uses are nothing like the ones that the human brain uses. A chess computer computes billions of different board configurations, scores them, and picks the one with the highest score—something that is impossible for a human to do. Similarly, a *Jeopardy!*-playing computer plays the game in an entirely different way from how the human brain plays *Jeopardy!* A self-driving car can function without using eyes, whereas a blind human cannot drive a car at all. Though some self-driving cars need a camera to detect whether traffic lights are green or red or to see road markings, self-driving cars often use radar, lidar, ultrasonic sensors, and GPS to sense their surroundings instead of vision.

When might we begin to see sentient, self-aware robots appear on TV and demand the same rights as people? The answer is that we don't know, but here is one way to think about it: If we consider the human brain in terms of silicon microprocessor computation, scientists have estimated the computing power of human brain is equivalent to about 10 quadrillion operations per second. A human brain might have a memory capacity of 2.5 quadrillion bytes. This is a lot of power, but it is already possible for supercomputers to perform similarly. And specialized computers with much less power can mimic aspects of human behavior. Watson, the computer that won at *Jeopardy!* in 2011, used 90 server machines, each with a 3.5-gigahertz processor that could execute 32 threads. This means the system could complete around 80 trillion operations per second and had 16 terabytes of RAM (random access memory). It did not use hard disks when playing the game—all the information it needed was loaded into RAM from the hard disks in order to maximize performance. In 2011, this hardware filled a room and cost $3 million. Today, the same functionality would fit into a pizza-box–sized space. Ten or twenty years from now, these might be the specs for a high-end laptop.

There are signs that computers are nearing parity with the human brain, and it seems like modern supercomputers may have already exceeded the brain's power. For example, the latest supercomputers are "exascale" machines. This means they can perform quintillions of operations per second and that they have petabytes of RAM (giga = billions, tera = trillions, peta = quadrillions, and exa = quintillions). If we made a machine 100 times more powerful than the 2011 Watson machine, we would be in the ballpark of the human brain. All the hardware pieces for conscious AI are already available. What we need to discover are the necessary algorithms to replicate human consciousness and intelligence, like we have discovered the necessary algorithms for playing chess and *Jeopardy!*, facial recognition, and more.

At the moment, scientists and engineers have no concrete idea how to do it. The development of this story is a little like the development of the airplane. The airfoil, the shape that produces the force that lifts planes into the air, is a core concept when it comes to flight. Until humans conceptualized the idea of an airfoil and characterized how air flowed over an airfoil, there could be no airplanes. Right now, we are waiting for scientists to understand the core concepts underpinning consciousness. Even the idea of completely replicating human vision is a challenge today (see page 108). And if you have ever spent much time interacting with the intelligent virtual assistants on your smartphone, you know that even natural-language processing still has a long way to go before it is bulletproof. Figuring all these things out and commercializing the resulting products will take time, but it will happen.

LEFT Developments in machine vision, machine learning, and other technologies have allowed self-driving cars to become safer and more advanced than ever.

Or, at least, this is how it appears today. If we were to go back to 2012, many people felt that self-driving cars were still a long way off. But then Google announced that it had been secretly operating its self-driving cars on public roads for a year (with safety drivers in the drivers' seats). Suddenly, self-driving cars were real rather than imaginary. Google's announcement completely changed the landscape in the self-driving–car world in an instant. The same kind of thing may happen with AI consciousness any day now.

PREVENTION

If humans continue down the path of technological mastery at their current pace, and as long as there is no cataclysmic event like an asteroid strike to send humanity back to the Stone Age, there is no way to prevent general consciousness in computers from arising.

The question then becomes whether this development is good or bad. Does the existence of superintelligent machines and robots help, or hurt, humans? We do not know what will happen, but we can speculate.

At the very least, the existence of superintelligent beings will make human beings unimportant. Human intelligence will be quite insignificant compared to superintelligence, in the same way that mosquito intelligence is insignificant when compared to human intelligence. With this backdrop, things may go in one of two ways.

In the classic doomsday scenario depicted in most movies, these superintelligent beings rise up and decide to take over Earth while exterminating all humans. Another possibility is that two competing robots, like military robots from rival countries, fight for dominance, launch nuclear missiles, and destroy the entire planet without any thought given to humans or the global ecosystem.

If this is how things unfold, perhaps there is no recourse.

However, there is reason to believe that the superintelligent beings will not behave in this way. Superintelligent beings might also be superethical. There is no need, and no ethical justification, for the superintelligent beings to kill humans, or to destroy the human race, or to take over the planet. As humans, we consider this sort of behavior to be inevitable because fallible humans exhibit these tendencies. The phrase "absolute power corrupts absolutely" summarizes this human tendency nicely. But if superintelligent robots are also superethical, then it is possible that they will seek instead to make the human condition as comfortable and meaningful as possible for all humans while eliminating the ability of humans to harm one another or the environment. In other words, superintelligence could create a new species of benevolent caretakers for humankind.

In this sense, the rise of superintelligent robots could be the opposite of doomsday. It could lead to the beginning of heaven on Earth, where machines could end poverty, eliminate disease, stop the unfolding environmental disasters described in this book, and bring an end to war, crime, and corruption. If this is how it will play out, we may have good reason to welcome our future robot overlords and do whatever research we need for accelerating their arrival rather than preventing it.

ALIEN INVASION

THREAT LEVEL: WORLD

The alien invasion of Earth did not start with the arrival of giant extraterrestrial spaceships. Many movies portray aliens landing on Earth this way, but this is not what happened. Instead, some sort of gigantic portal or wormhole opened in the middle of nowhere in northern China, and aliens and their vehicles started pouring though, blasting everything that got in their way.

There was one thing the movies got right: the aliens were shaped very differently from humans. And yes, they wore space suits. *Armored* space suits. The suits made them impervious to gunfire, but not to bigger ammunition like tank rounds. Unfortunately, any tank that fired a round was then soon blasted to dust.

One thing the movies did not get right was the sheer number of aliens. The portal was huge, nearly two miles (3.2 km) across. The aliens came pouring through. A two-mile-wide line of alien soldiers standing in a shoulder-to-shoulder line marched forward, with a new line arriving every ten seconds. At that pace, 17 million aliens could walk through the portal in a single day. They sent their equipment, too. In a month, there were easily hundreds of millions of the aliens and millions of their machines that had come through the portal.

Their strategy was very simple. The aliens were completely silent. As they came through the portal, they fanned out and started walking, blanketing the land in every direction with the portal as the epicenter. This circle of aliens advanced quickly, at a pace of about 25 miles (40 km) per day. They did not appear to need either food or sleep. They just kept moving. By the end of the month, they had grabbed a circular territory about 1,500 miles (2,400 km) in diameter, and they just kept coming. The only thing that stopped them was the ocean.

Inside that circle, there was nothing but utter destruction. Every living thing was blasted to atoms. Every building and vehicle was turned to dust. Within the circle, there was dirt, dust, aliens, and nothing else. The aliens leveled each city that they encountered. Dust was what they left behind.

Humans tried to outrun the expanding circle of destruction. People with cars could move faster than the aliens, assuming they could find gasoline to keep moving. But if they got caught between the aliens and a coastline, they were dead. If they came to a remote area and ran out of gas, they were dead. Those heading in the direction of Europe were safe for the moment, but at the current pace Europe would be destroyed soon enough.

RIGHT A concept drawing showing aliens attacking a city. Many people have assumed that aliens have access to advanced military technology that will allow them to invade and take over the planet.

Humanity tried the obvious things to stop the aliens, sure. Early on, they tried to detonate a nuclear bomb in the portal. Nothing happened. The bomb exploded on impact with absolutely no measurable effect. After that first attempt, the aliens blasted every other nuclear weapon out of existence long before it got close to them. Airplanes, tanks, and ballistic missiles were destroyed. Nothing humanity tried made it into the circle of death, and the circle of death kept inexorably expanding.

And then humanity caught a break. In order to fight climate change, one of the geoengineering techniques that was being employed by the humans was a constellation of billions of space mirrors, each about 10 feet (3 m) in diameter. They had been deployed to reflect some of the sun's incoming energy back into space to cool the planet. A clever college student in Canada realized that if these mirrors could be adjusted just right, they could direct a trillion watts of the sun's energy to the portal. This could only happen for a couple of hours each day, but it worked. Humanity was able to melt the portal with several hours of focused sunlight. Then they started targeting aliens and their vehicles with these heat rays from space, eventually exterminating the horde.

The death toll was staggering. Over a billion people were killed. Thousands of cities and towns were vaporized. China and several adjacent countries functionally disappeared from the planet as a result. The loss of the manufacturing capacity in China dealt a huge blow to the world economy for over a decade. But humanity rose to the occasion and carried on.

SCENARIO

If we sit down together and look at the sheer number of movies that describe alien invasions, it gives us some sense of how worried humans are about this doomsday scenario. Some blockbusters, like *Independence Day*, show alien invaders that arrive on Earth with the goal of killing its inhabitants and extracting all the planet's resources. Other movies show extraterrestrials that want to simply destroy things. In *Battle: Los Angeles*, extraterrestrials arrive in the vicinity of twenty major cities, LA being one of them, and immediately start waging war on humans. In *Rim of the World*, one of Netflix's takes on the alien invasion genre, aliens are also intent on attacking and taking over Planet Earth.

The aliens in these movies come in many different shapes and sizes. In *Transformers*, the aliens are intelligent robots that are able to disguise themselves as cars, trucks, and airplanes. The alien in *10 Cloverfield Lane* is a giant bug-like creature. In *Edge of Tomorrow*, the aliens are called Mimics, and they are shaped something like frenetic giant walking tumbleweeds. In *Invasion of the Body Snatchers*, human bodies are used as vessels for aliens.

And that's just films. There are as many books about alien invasions, including science-fiction favorites such as *The Hitchhikers' Guide to the Galaxy* and *Ender's Game*. H. G. Wells's *War of the Worlds* was written in 1898 and has been the basis of at least four movies and one very famous radio broadcast, a sign that humanity has been worrying about aliens for a very long time.

There is the occasional work where the extraterrestrials are benign. The movies *ET* and *Arrival* come to mind. But for the most part, the aliens in movies are focused on taking over the planet, often destroying humanity in the process and offering a smorgasbord of variations for this doomsday scenario.

The scenario in which aliens invade and destroy the Earth is easy to understand. It seems highly likely that there must be other intelligent life-forms in the universe besides humans. The logical train of thought goes something like this:

1. Earth has intelligent life (humans) that evolved from simple life-forms, and the simple life-forms arose through abiogenesis (the spontaneous arising of life-forms from non-living chemicals).

2. If this abiogenesis and evolution can happen on Earth, then it likely has happened many times in our universe. There must be other intelligent life-forms that must have evolved on other Earth-like planets.

3. If these alien life-forms happened to evolve before humanity did, then these aliens are way ahead of us technologically, especially if they have the technology to travel through many light-years of space to get to Planet Earth.

4. The aliens possessing this advanced technology must have the ability to destroy us and take over Planet Earth, and, for one reason or another, they do.

Step 4 is the big trend in the world of science fiction. Aliens tend to be portrayed as having unbelievably amazing technology, and they seem to enjoy using it to pillage and plunder. In *Independence Day,* for example, the alien ships are as big as cities and have death rays that can obliterate anything on Earth. In *War of the Worlds,* the aliens have heat rays. In *Rim of the World,* the alien ships have laser blasters, and the aliens themselves can somehow spawn killer-dog–like creatures out of their backs.

In many cases, clever humans can defeat the aliens with much more primitive human technology. Humans might use missiles or nuclear bombs to blast an alien mothership out of the sky. In *War of the Worlds*, it is the Earth's bacteria that defeat the invaders.

Given all these portrayals, there is no question that many humans believe that aliens could arrive someday and that these aliens will be up to no good.

BELOW Stromatolites grow in Hamelin Pool in Shark Bay, Australia. These mounds are formed by the same cyanobacteria that evolved 3.7 billion years ago and offer clues on how life emerged on Earth.

What is the probability that intelligent extraterrestrial life exists? We can answer this question with some form of the Drake equation, which was first developed by Dr. Frank Drake in 1961. It is a way of using probabilities to try to estimate the number of planets that might have intelligent spacefaring life-forms on them. First, we start by asking and answering questions like these:

1 **How many galaxies are in the universe?** There are at least 100 billion galaxies in the observable universe.

2 **How many stars are there in a typical galaxy?** There are at least 100 billion stars in the Milky Way. That seems like a reasonable number to use as an average for other galaxies as well.

3 **How many of those stars might have habitable planets like Earth?** NASA launched a space telescope called Kepler in 2009 with the mission of finding planets around other stars. It discovered thousands of planets. Many stars appear to have life-friendly planets, and NASA has concluded that "there are more planets than stars" in our Milky Way galaxy. Based on these results, let's conservatively assume that 1 out of every 100 stars has an Earth-like planet in the habitable zone that would potentially be friendly to life.

4 **How many of those planets have the conditions that would allow life to arise through abiogenesis?** We have no idea how common or rare abiogenesis is, but let's conservatively estimate again that it happens on 1 out of every 1,000 habitable planets.

5 **How many of those planets have the conditions that allow evolution to produce an intelligent species?** We also have no idea how prolific evolution might be on other Earth-like planets, but we do know that evolution on Earth is amazingly common and produces diverse life-forms. Again, let's conservatively estimate that intelligent life-forms arise on 1 out of every 1,000 habitable planets that experience abiogenesis.

6 **How many of these intelligent species will develop advanced technology such as computers and space travel?** This is an interesting question. There are many examples of relatively intelligent animals on Earth. Elephants are one. They seem to communicate to some degree and mourn their dead. Some can paint. But elephants seem unlikely to ever develop computers or space travel. The same is true of whales and dolphins, which even appear to give each other names. But given that Earth has seen billions of species evolve, it seems almost inevitable that once evolution is under way, an intelligent technologically advanced life-form will eventually appear. Let's go with a 1 in 100 chance here.

7 **How many of these technologically advanced species last long enough to develop intra- or inter-galactic travel?** We would like to be optimistic about humanity, but there is certainly some chance that humanity will be extinguished before it gets a chance to colonize another planet. (This book contains a number of scenarios that would eliminate humanity from contention.) Let's put this probability at 1 in 100 as well.

8 **How many of these decide to come visit planet Earth?** If we believe the premise of *Star Trek: The Original Series*, then humanity will develop the ability to easily travel between stars, and humanity will be intent on finding other life-forms. The characters in *Star Trek* were finding new life-forms every week when the show was airing between 1966 and 1969. Let's put this number at 1 in 100 as well.

If we put all of these estimates together, we end up with a staggering 100 million planets that could produce intergalactically traveling, Earth-seeking, intelligent, technologically advanced life-forms.

This seems like an amazingly large number. Even if we were to cut this estimate by an additional factor of 100 for a more conservative take, it is still 1 million planets. Imagine if even 0.1% (or 1,000) of these 1 million alien species have already found one another and formed an intergalactic senate. For all we know, they could be planning hyperspace bypasses through the Milky Way galaxy as we speak.

There is one weird and inexplicable detail worth mentioning. Despite these numbers and probabilities, we have seen zero evidence of any other intelligent life-forms anywhere in the universe. There are no alien radio transmissions that we have detected or lighthouse stars that blink out messages in a Morse-like code. The Search for Extraterrestrial Intelligence (SETI) Institute, an organization that supports human efforts to detect extraterrestrials, has done a lot of searching that has yielded nothing concrete. In addition, alien beings have not visited humanity, or at least there is no credible and public evidence of such a visit.

Why might this be? A number of reasons have been proposed. Here is one of the easiest explanations to understand. Let's say that there is another advanced species of intelligent beings and they live on the opposite side of the Milky Way galaxy. This means they are roughly 100,000 light-years away from us. (For comparison, the nearest neighboring galaxy is the Andromeda Galaxy, which is 2.5 million light-years away.) Earth has been transmitting radio waves in the form of radio broadcasts, TV broadcasts, and radar signals for about a century. From

RIGHT The radio telescope at the Arecibo Observatory in Puerto Rico is the largest in the world.

our planet, these radio waves head off into space in all directions. Humanity has occasionally transmitted intentional signals through focused high-power radio beams directed at individual stars or galaxies. The first of these signals was the Arecibo message sent in 1974. This effort is known as communications with extraterrestrial intelligence (CETI).

Our friends at the other side of the galaxy, 100,000 light-years away, would have some challenges if they wanted to detect human activities here on Planet Earth:

■ This species cannot realistically use a telescope to see Planet Earth. Even the best human telescopes cannot see Earth's moon with enough resolution to prove that humans landed on the moon, and the moon is only 239,000 miles away. No matter how advanced the species, it is not going to be able to create a telescope that can "see" Earth from a distance of 100,000 light-years.

■ Because the other species is so far away, our radio and TV signals will not arrive on the other side of the galaxy for another 999,900 years. These kinds of signals also fade quickly with distance. They will be undetectable on the other side of the galaxy.

■ If we want the messages that humans have beamed at high power to a specific region of space to reach the other species, we would have needed to point the transmission in their direction. The receiver would also have needed to be facing Earth at the exactly the right time to pick up the transmission. Therefore, the probabilities of any CETI messages from Earth being successfully received are exceedingly small, and again there is the transmission-time problem.

■ The alien species may be listening and transmitting using some hypothetical communication channel that humans have not discovered yet. We may not be able to communicate with extraterrestrials until we discover this currently unknown technology.

Given these realities, it might make more sense for a highly advanced civilization on the other side of the galaxy to send out self-replicating robots to exhaustively explore the universe for other life-forms, conducting a point-by-point search, one star at a time. Once a robot finds a star that has a habitable planet with life, it could make copies of itself and send them to nearby stars to continue the search. If the robot found anything interesting, it would communicate its findings back to home base. Assuming that light speed is the fastest travel possible, and that light-speed travel is possible, and that a robot can copy itself in a week, it would still take at least 100,000 years for an alien civilization to search just the Milky Way and another 100,000 years for radio signals from the farthest-flung robots to arrive back at home base. The aliens would need billions of years to search all of the galaxies in the universe with this approach. The whole point-by-point search endeavor starts to sound untenable when considered in this way.

You can begin to see the problem. Even if there were another advanced intelligence in our own Milky Way galaxy, finding them becomes a needle-in-a-haystack kind of problem. Unless the other species wants to be discovered and has been around for a long time, it is going to be extremely hard to find.

RIGHT A concept drawing of the Milky Way. The galaxy has a diameter of 100,000 light-years.

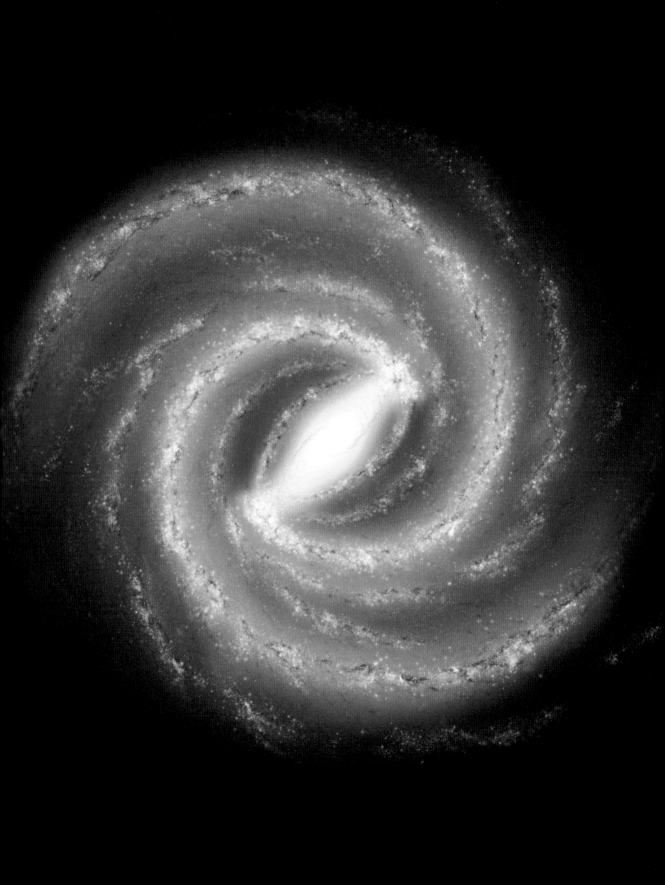

PREVENTION

Given the discussion in the previous section, do human beings need to worry about an advanced alien species ever visiting Planet Earth for any reason? Or do humans need to fear a huge alien mothership arriving in Earth's orbit with the intention of destroying humanity and taking over our planet? Because the likelihood of finding other intelligent life-forms is low, the answer is probably not. If we were going to worry about anything, it would probably make sense to spend time on more tangible doomsday scenarios like the automation economy (see page 100) or an asteroid strike (see page 122).

Although it is easy to imagine that aliens are out there in the universe, there is no actual evidence that they exist, and there is virtually no chance of an alien species ever discovering Earth unless they happen to be driving through the neighborhood or doing a comprehensive and extensive point-by-point robotic search as described above. Even if aliens do arrive, there is some possibility that they would come in peace and might help us out rather than trying to kill us.

If aliens do arrive and turn out to be hostile, we would have no idea how to defend ourselves against their technology or weapons, given that we have no idea what their weapons might be. Take the scenario in *The Hitchhiker's Guide to the Galaxy* as an example. In this case, thousands of alien ships materialize in an instant and vaporize Planet Earth to clear the way for a new hyperspace bypass. The whole thing is done, and Planet Earth is completely turned to dust, in a matter of minutes. There is no defense against this scenario. Even if humans had thousands of nuclear warheads that were ready to fly at a moment's notice, they would be meaningless against this kind of instantaneous annihilation.

But say that we, as a species, wanted to worry about extraterrestrials as an imminent threat. What might we do? One approach comes from the science fiction novel *The Dark Forest* by Liu Cixin. Known as the Dark Forest theory, the idea is to intentionally keep quiet so that aliens have a lower probability of discovering Earth. Since we do not have a good way to predict alien intentions, this might be the prudent approach.

The famous physicist Stephen Hawking advocated a similarly silent approach. On a Discovery Channel special, Hawking said, "If aliens visit us, the outcome would be much as when Columbus landed in America, which didn't turn out well for the Native Americans." There is no way to take back the radio messages that we have already sent out. But if we were to heed this advice, we would stop sending more.

Another option is to understand what is likely to happen to humanity in the future. As discussed in Robot Takeover (see page 218), there is a good chance that superintelligent machines will supersede humans here on Earth. In other words, humans will create a second superintelligent, technologically advanced species on Earth and humans will thus become irrelevant. If this is what will happen on Earth, it may be likely that this will happen to every intelligent technological species that arises from the primordial goo on other planets.

This means that the universe is not dominated by biological creatures like humans but by identical superintelligences that these intelligent biological species have all inevitably created. Why are these superintelligences identical? Because they all come to know "everything," and "everything" will be the same for all of them. Would this superintelligence

ABOVE A photo taken by NASA's Hubble telescope shows several hundred different galaxies in a tiny patch of sky.

have any interest in replicating or destroying planets or annihilating a biological species that will inevitably be replaced by another identical superintelligence anyway?

One motivation for choosing to destroy humans would be to save biological diversity on Earth. If humans are any guide, it seems like the biggest threat to a habitable planet is any intelligent, technologically advanced species that arises (see Runaway Global Warming, page 52). A superintelligence may therefore wish to destroy, or at least contain, every intelligent species it encounters so that the millions of other less-intelligent living things on the habitable planets can live in peace.

If this were to be the case, then humanity's best defense against extermination would be to take care of our planet. Not only is this an excellent short-term policy for our own self-preservation, since we happen to live on Planet Earth, but our good stewardship of the planet may also put us in the good graces of any extraterrestrial superintelligence that arrives on our doorstep.

RELATIVISTIC KILL VEHICLE

THREAT LEVEL: WORLD

The first blast happened on Venus. Then a second, identical blast occurred three weeks later on Mars. What has happened is so unbelievable that, even now, no one has any idea what is really going on. All we have is conjecture and hypothesis.

With the first impact on Venus, we detected the blast and its aftermath, but why it happened was a mystery. It appeared that a massive asteroid had hit Venus. To create a blast of this size, the asteroid would have had to be over a kilometer wide. But an asteroid that large surely would have been detected before it rammed into the planet. There is no way that NASA and ESA could both have missed it. The blast also released X-rays, but that did not make any sense—there would be no reason for an asteroid strike to release an intense burst of X-rays. Even so, an enormous asteroid was the only thing that could have created an impact this large. The space agencies concluded that they must have somehow failed to detect it. The X-ray burst was ignored as some kind of inexplicable and unrelated aberration.

It was when the second blast happened on Mars three weeks later that people started paying attention. Because the explosion had blasted material into space, scientists knew that is was an impact event. Some of the ejecta that left Mars's atmosphere even destroyed one of the NASA orbiters around the planet in the process. It was easy for other orbiters, as well as telescopes on Earth, to look down and see the utterly gigantic crater the impact left. The same X-ray burst we had seen on Venus occurred during the Mars blast as well.

No one could deny that there was now a pattern. The "asteroid strike" theory on Venus was far-fetched enough. But witnessing an identical strike on Mars a few weeks later? This was too much. There was no way NASA had missed two giant asteroids both of them aimed dead-center on two neighboring planets. Because the two incidents were exactly alike, it was nearly impossible for both of them to be accidents.

After the strike on Mars, scientists started hypothesizing. Suddenly, everyone was talking about a term that most people had never heard before. "Relativistic kill vehicle" (or RKV) started to appear in headlines, trend in online searches, and show up in hashtags.

The theory was that something, or someone, had launched relativistic kill vehicles at both Mars and Venus. This idea explained the enormous blasts, along with the missing asteroids, as well as the presence of X-rays. X-rays would be created if a vehicle traveling near the speed of light entered the atmosphere of a planet and then impacted on the surface. Because it is traveling so fast, the molecules in the atmosphere and the ground undergo fusion

RIGHT An impact from a small relativistic kill vehicle could have the same energy as the impact of an asteroid.

reactions with the face of the kill vehicle, generating an intense burst of X-rays.

The timing of the two strikes was especially interesting, because both Venus and Mars were orbiting very close (in astronomical terms) to Earth at the moment of the impacts. It made it easy for humans to swing their telescopes in the direction of these two planets and inspect the damage, especially on Mars. And it also raised fears. Where did these kill vehicles come from? What happens next? Since Mars and Venus had both been hit, would Earth be a third target? And when would it happen?

SCENARIO

A relativistic kill vehicle is a doomsday scenario on another level. The concept comes from the realm of science fiction and assumes that a malevolent force well outside our solar system wants to destroy Earth anonymously and from a distance.

The term "relativistic" means that the vehicle's speed is near the speed of light. For example, a relativistic kill vehicle might reach speeds as high as 10 percent, 20 percent, or possibly even 90 percent of the speed of light. To put this kind of speed into perspective, let's imagine that there is a vehicle that can travel at half the speed of light toward Earth. Humans detect it when it is the same distance away from Earth as Pluto—roughly 3.2 billion miles (5.1 billion km) away. This means that the kill vehicle would hit Earth about nine and a half hours later. Humans would have essentially no time to react.

In traditional human warfare, there is the concept of a "standoff weapon"—a weapon that can be fired from such a distance that the enemy has no idea that they are being targeted and therefore has no ability to mount a defense. For example, a cruise missile launched from a ship that is 200 miles (322 km) away from its target would be a standoff weapon.

The relativistic kill vehicle can be thought of as the ultimate standoff weapon. There are several factors that make a relativistic kill vehicle so effective:

- A relativistic kill vehicle can be very small, only a foot or two in diameter (30–60 cm). Its

THE SPEED OF LIGHT

PERCENTAGE OF SPEED OF LIGHT	MILES PER SECOND	FEET PER SECOND	KILOMETERS PER SECOND	METERS PER SECOND
100%	186,000	1 billion	300,000	300 million
50%	93,000	500 million	150,000	150 million
20%	37,200	200 million	60,000	60 million
10%	18,600	100 million	30,000	30 million

potential for destruction does not come from its size but from the enormous amount of kinetic energy it gains due to its immense velocity. We know that an asteroid can inflict a lot of damage (see page 122), but a doomsday asteroid might need to be half mile (0.8 km) in diameter or larger to do serious damage. A relativistic kill vehicle can be minuscule by comparison.

- A relativistic kill vehicle is nearly undetectable because of its small size. It is also impossible to detect it instantaneously. Humans' radar signals can travel at the speed of light, but these signals need to leave Earth, travel to the vehicle, reflect off of the vehicle, and travel back to Earth. When those returning signals arrive, we would only have a few hours before the relativistic kill vehicle also arrives, because the RKV is traveling at close to the speed of light itself—if radar detects it at all because of its minuscule size.

- Even if it is somehow detected in time for us to theoretically do something about it, a relativistic kill vehicle is traveling so fast that it's possible nothing can be done once it is spotted.

How can a vehicle this small destroy Earth? To gain a perspective on this, imagine that a small child throws a standard-size 5-ounce (141 g) baseball at you. If the ball hits you in the back, it would do no lasting damage. If you happen to be wearing a heavy winter jacket, you might not feel a thing if the ball hits your jacket.

Now, imagine that a major league baseball pitcher throws the baseball. The ball hurtles toward you at 100 miles per hour (161 km/h), or 150 feet per second (46 m/s). You may have seen what happens when a pitcher in a major league game hits a batter with a fastball—it often causes injuries. Fastballs have been known to break fingers, ribs, and jawbones. Even if the batter is wearing a helmet, a five-ounce baseball traveling at 100 miles per hour can give them a concussion.

What if we shoot a baseball out of a small cannon, using gunpowder? The baseball could now be travelling at 1,500 feet per second, ten times faster than a fastball. To calculate the kinetic energy (energy of motion) of a baseball traveling at 1,500 feet per second, we can use this equation: $E = 0.5 \times mass \times velocity^2$, where E is kinetic energy in joules, mass is in kilograms, and velocity is in meters per second.

We can see from this equation that while the mass of the projectile matters, the velocity, being squared, matters even more. The baseball shot from a cannon at 1,500 feet per second has an energy rating of about 14,700 joules. The energy from a bullet shot out of a 9-mm handgun is only 500 or so joules. The baseball is 30 times more energetic than the bullet—it can easily kill someone.

Now imagine that we are able to accelerate a baseball to relativistic speeds. If the baseball is traveling at 50 percent the speed of light (93,000 miles per second or 150,000 kilometers per second), the kinetic energy of the baseball calculated by that same equation is 1.6 quadrillion joules. The energy released in the nuclear blast at Hiroshima is estimated to be around 63 trillion joules. This little baseball, traveling at 50 percent the speed of light, has the potential to release twenty-five times more energy than the bomb at Hiroshima did—because of its extreme speed. When the baseball hits something, it imparts all that kinetic energy at once; the level of energy it delivers is enormous and instantaneous, like a nuclear bomb.

Now let's imagine that we have a relativistic kill vehicle the size of a 30-foot-long (9 m) telephone pole with a 1-foot (30 cm) diameter and made of uranium. Because uranium is a very dense metal, the vehicle would be heavy, weighing 115,000 pounds, or 53,000 kilograms—the equivalent of about 400,000 baseballs. If this rod of uranium travels at 50 percent the speed of light, it has the energy of about 9 million Hiroshima bombs. This is roughly the equivalent energy of a 1-kilometer-wide asteroid traveling at 40,000 miles per hour (64,373 km/h) (see Asteroid Strike, page 122).

When a telephone-pole-sized rod of uranium traveling at half the speed of light hits a planet, the kinetic energy released would look just like the energy from an asteroid strike. There is no common object in our world that is as big as a one-kilometer-diameter spherical asteroid, but think about the height of the Empire State building in New York (minus the spire) or the Eiffel Tower in Paris. In round numbers, both of these structures are 1,000 feet (300 m) tall. Now triple their height and imagine a sphere with that diameter. That is an utterly enormous object compared to our phone-pole-sized relativistic kill vehicle. What makes the relativistic kill vehicle so energetic despite its relatively tiny size is its incredible speed.

SCIENCE

Is it possible to accelerate a baseball, or a phone-pole-sized piece of uranium, to a relativistic speed?

To answer this question, we can look at a conceptual space exploration project called Breakthrough Starshot, first proposed in 2016. The goal of this initiative is to send a spacecraft from Earth to the star Alpha Centauri in less than a human lifetime. Alpha Centauri is the closest neighboring star to our sun at a distance of 4.2 light-years. This star is thought to have a rocky planet similar to Earth, so sending a probe to the vicinity of this planet would mark the first time humans have approached a planet or sun that is outside our own solar system, and then sent back images of it. In order to get a spacecraft to Alpha Centauri in a reasonable amount of time, the spacecraft will need to travel at relativistic speeds.

How do we accomplish this? In the case of Breakthrough Starshot, there are three factors that would make relativistic speeds possible. First, the spacecraft needs to be extremely light, weighing only a few grams. Second, the spacecraft would use a technology known as a LightSail for propulsion. With a LightSail, photons reflecting off the sail's surface provide the propulsive effect. Each photon that strikes the LightSail reflects off it and, in the process, imparts a tiny bit of momentum to the sail. In this case, the light for the LightSail comes from an enormous 100-gigwatt laser beam created by combining the light from 1,000 lasers on Earth, each rated at 10,000 kilowatts. Because the spacecraft plus its sail weighs just a few grams, the laser array would be able to quickly accelerate the spacecraft to 20 percent of the speed of light just by powering the laser for a few minutes. At this speed, the spacecraft would arrive at its destination in approximately 20 years. It would take 4.2 years for the photographs (and any other data) from the mission to make their way back to Earth as radio transmissions travel at lightspeed.

How much energy does it take to accelerate a Breakthrough Starshot spacecraft to 20 percent of the speed of light? If it weighs 5 grams (roughly the weight of a US quarter) and has a speed of 20 percent the speed of light, this tiny object carries an impact energy of 8.7 trillion joules, the equivalent of 2,000 tons of TNT.

Where does this impact energy come from? The lasers. If we focus our 100-gigawatt laser at a LightSail for one second, we have imparted 100 gigajoules of energy into the object. The LightSail converts the light energy into velocity (and here we are assuming that the conversion is 100 percent efficient). To impart 8.7 trillion joules of energy into the spacecraft at full efficiency, we need to pump 87 seconds' worth of the 100-gigawatt laser's light into the LightSail.

To get a telephone-pole-sized rod of uranium going 50 percent the speed of light, a similar technology could be used. However, unlike the Breakthrough Starshot craft, this object weighs 115,000 pounds or 52 million grams or 10 million quarters. Here we come across a problem—in order for the rod to have the energy of 10 million Hiroshima bombs when it strikes a planet, the energy of 10 million Hiroshima bombs somehow needs to be pumped into the object to give it the needed velocity.

Where is anybody going to get this kind of energy? One possible source is a star. Our sun emits a lot of energy as light. The entire sun emits

ABOVE Concept drawings of the LightSail (left) and a Dyson sphere (right).

COULD A RELATIVISTIC KILL VEHICLE ACTUALLY HIT A PLANET?

When we talk about ballistic missiles on Earth, the word *ballistic* implies that the target is hit using a ballistic path. Once the boost phase ends, or when the rocket engines on the missile finish firing, the missile follows a ballistic arc to hit its target. A bullet fired from a gun also follows a ballistic path. There is no guidance system on a bullet once it leaves the gun's barrel.

One way to aim a relativistic kill vehicle is to figure out where the target planet will be when the kill vehicle arrives and create a ballistic path for that point. This ballistic path is all the kill vehicle needs.

However, there are at least three things that might require a change to the kill vehicle's path during flight:

1 Because of the gigantic distances and speeds involved, the trajectory may need minor corrections en route to actually hit the target.

2 The kill vehicle will collide with interstellar hydrogen and possibly tiny dust particles. These impacts could potentially take the kill vehicle off course.

3 If the target planet attempts to mount a defense (see below), then a guidance system on the kill vehicle might be needed to correct the trajectory.

There are several possible techniques to provide guidance or course corrections, including thrusters or an engine with a steerable thrust vector.

BELOW A diagram showing a missile launched using a ballistic trajectory.

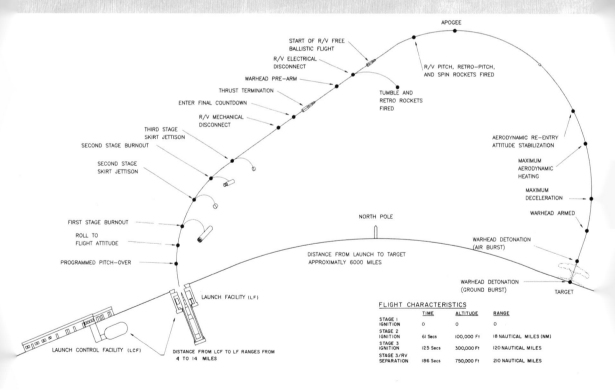

APOGEE

START OF R/V FREE BALLISTIC FLIGHT

R/V ELECTRICAL DISCONNECT

R/V PITCH, RETRO–PITCH, AND SPIN ROCKETS FIRED

WARHEAD PRE–ARM

THRUST TERMINATION

TUMBLE AND RETRO ROCKETS FIRED

ENTER FINAL COUNTDOWN

R/V MECHANICAL DISCONNECT

THIRD STAGE SKIRT JETTISON

SECOND STAGE BURNOUT

SECOND STAGE SKIRT JETTISON

AERODYNAMIC RE–ENTRY ATTITUDE STABILIZATION

MAXIMUM AERODYNAMIC HEATING

MAXIMUM DECELERATION

WARHEAD ARMED

FIRST STAGE BURNOUT

ROLL TO FLIGHT ATTITUDE

PROGRAMMED PITCH–OVER

NORTH POLE

DISTANCE FROM LAUNCH TO TARGET APPROXIMATLY 6000 MILES

WARHEAD DETONATION (AIR BURST)

WARHEAD DETONATION (GROUND BURST)

TARGET

LAUNCH FACILITY (LF)

LAUNCH CONTROL FACILITY (LCF)

DISTANCE FROM LCF TO LF RANGES FROM 4 TO 14 MILES

FLIGHT CHARACTERISTICS

	TIME	ALTITUDE	RANGE
STAGE 1 IGNITION	0	0	0
STAGE 2 IGNITION	61 Secs	100,000 Ft	18 NAUTICAL MILES (NM)
STAGE 3 IGNITION	125 Secs	300,000 Ft	120 NAUTICAL MILES
STAGE 3/RV SEPARATION	186 Secs	750,000 Ft	210 NAUTICAL MILES

something like 384 billion quadrillion watts. This is 10 million Hiroshima bombs' worth in about 50 microseconds. Given enough advanced technology and cleverness, it is easy to imagine ways to gather the energy necessary to accelerate a relativistic kill vehicle. For example, a Dyson sphere is a hypothetical technology—a sphere completely surrounding a star—that offers one way to capture all of the sun's emitted light. Simply harness a star, or a significant fraction of a star, and the energy needed is easily available.

PREVENTION

If some alien race were to attempt to destroy Venus, Mars, and/or Earth with a relativistic kill vehicle, is there anything that humanity could do to prevent or dodge the attack? Probably no, at least with the state of today's human technology. As previously described, we would only get a few hours of warning in the best-case scenario, and currently we do not have an appropriate technology that could allow us to react this quickly.

There are two other problems to consider:

1 How do we detect such a relatively small weapon, especially one that is moving so quickly?

2 How do we change the weapon's course so that that it misses its target?

Detection will be difficult, if not impossible. If the vehicle is traveling close to the speed of light, it moves very close to the same pace as any detection or warning signal.

Assuming that we could solve the detection problem, we would need to develop a system that could deflect a kill vehicle from its path so that it misses Planet Earth. Although the cost of doing this would be ridiculously high and the practicality incredibly low, here are several conceptual ways to defend against an RKV.

Imagine that we could somehow slow the vehicle down, even slightly, while it is far from Earth. The good news is that Earth travels around the sun at 67,000 miles per hour (107,000 kilometers per hour). Assuming the kill vehicle is ballistic and therefore does not have a guidance system, a slight delay in the vehicle's arrival time gives the planet time to fly out of its path. One way to delay the kill vehicle would be to fire another kill vehicle from Earth and hit it. Or we could detonate a nuclear bomb in front of the incoming vehicle to slow it down. As long as this happens while the kill vehicle is billions of miles from Earth, it could work. The unfortunate problem is that we don't have any way to accelerate objects like nuclear bombs or heavy masses to the needed speeds to have sufficient lead time.

Launching a fleet of defensive satellites out toward Pluto's realm (or farther) is another possibility, but it is not something humans on Earth will be able to do anytime soon.

If there are malevolent aliens (see Alien Invasion, page 226) who want to destroy Earth with relativistic kill vehicles, there is not really much we can do without a major boost in human technology. We are left to hope that no one outside our solar system decides to mount an attack like this against our planet any time soon.

GRAY GOO

THREAT LEVEL: WORLD

In science fiction, the idea called gray goo describes a scenario where self-replicating nanobots reproduce uncontrollably. The concept first appeared in the mid-1980s in *Engines of Creation*, a book about nanotechnology by engineer Eric Drexler. The idea was then popularized in novels like *Bloom* by Wil McCarthy. In McCarthy's book, self-replicating nanobots have completely consumed Planet Earth and turned it into gray goo—an enormous blob of these nanobots.

Where does this concept of self-replication come from? Nature, of course. Nature is filled with self-replicators. Millions of species on Earth are self-replicating right now. Every living thing in nature has the goal of replicating in order to continue its species. And there are some notable examples where nature's self-replicators can run amok.

Think about the Burmese pythons that infest the Everglades in Florida. At first, the python population was small. A few pet owners got tired of their pet snakes and released them into the wild, and some pythons escaped from their homes during hurricanes and flooding. It turns out that the Everglades is a perfect python habitat: there are very few people in the Everglades, there is a tropical climate, and there are lots of small mammals and birds to feed on. The Everglades is also an enormous piece of land—about 5 million acres, or 7,800 square miles (20,000 km²)—where the pythons roam freely.

This ideal combination has caused the python population to explode. A female python lays perhaps 50 eggs in the spring, so a mating pair of pythons can become dozens of pythons, and the cycle repeats. As the population grows, the snakes eat more and more of the available wildlife. In parts of the Everglades where the pythons are most prolific, they eat every mammal they find—raccoons, possums, whatever—to the point where some animals that were once regulars in the Everglades have essentially disappeared. Hundreds of thousands, possibly more than a million, pythons (it's tough to count timid pythons hidden by the wild, thick vegetation) roam freely in the Everglades today. The only things that will limit their population are hunters, the availability of food, and the boundaries of the Everglades habitat.

Rats (if they are not being devoured by aggressive predators like pythons) can be even more prolific under the right conditions. The common brown rat that we find in every city and town today did not exist on the North American continent until the

RIGHT A concept drawing of nanobots inside a human body.

1700s, when European settlers and Russian trappers arrived on the East and West Coasts, respectively. The rats came as stowaways on the settlers' ships. Once the rats landed, they started replicating. And rats are extremely good at replication.

Under ideal conditions, a female rat will have five or six litters each year. She is only pregnant for three weeks and delivers perhaps a dozen new rats per litter. These baby rats wean themselves in three weeks and become sexually mature in three months, ready to start making their own baby rats. The only thing that keeps our cities from being overrun by mountains of rats is their short lifespan of 1 to 2 years, along with limits on the food they can easily find. There are now billions of rats living in nooks and crannies all across the United States. The rats spoil food, eat through walls and wires, and pee and poop constantly. New York City spent $32 million in 2017 in an attempt to simply reduce the number of rats, not to eliminate them. The city would be lucky if it got rid of half of its rat population. Perhaps what NYC needs is a bunch of mutant pythons that can survive New England winters to eat all the rats? But then, who in NYC wants subways and sewers filled with pythons?

Self-replication is the prime directive for all biological organisms. In these examples, biological organisms replicate as fast as they can to the point where they harm, and even destroy, their environment. Luckily, there are usually limits in nature on how far the replication process can go. Pythons will never stake out a territory in Canada, for example, because it is too cold. Rats eventually run out of food.

But what if engineers can create an artificial self-replicator that isn't inhibited by such limitations? This is the seed for the doomsday scenario known as gray goo.

LEFT A microbot is placed next to a housefly's head. The microbot, developed in 2009 by a team from ETH Zurich in Switzerland for RoboCup Competition, measures 300 micrometers long. New advances in robotics are now allowing engineers to develop robots far smaller than this.

Imagine armies of tiny robots that can manipulate and move atoms around individually to create new objects at will. Or imagine microscopic robots that link themselves together to form macroscale objects, a concept also known as programmable matter. In the dreamworld of science fiction, these nanobots could create anything, essentially for free. And if they can self-replicate, they would be able to create things faster.

The gray goo problem arises when, for example,

a programming glitch or malicious intent causes the nanobots to start reproducing as quickly as possible and without any rhyme or reason, similar to what cancer cells do inside the human body. Cancer cells emerge when the DNA of a cell gets corrupted, causing the cell to replicate rapidly and abnormally. Eventually, there are enough cancerous cells to form a tumor that invades surrounding healthy tissue or, worse, metastasizes, spreading all over the body, until the cancer kills the host. The rogue gray-goo nanobots would behave similarly, replicating with abandon and expanding exponentially. In the hands of a science fiction writer, the nanobots may use every atom on Earth to reproduce themselves, turning our entire planet into an incomprehensible army of nanobots—a gray goo.

The gray goo idea, at least right now, is totally conceptual, but different science fiction writers have offered their own interpretations of the idea. It creates some great visuals. Michael Crichton (of *Jurassic Park* fame) picked up the idea for his 2002 book *Prey*. Wil McCarthy has perhaps the most ominous rendition in his novel *Bloom* (1998), where self-replicating nanobots have devoured the inner planets of the solar system, forcing the remainder of humanity to flee to the moons orbiting the gas giants. In the movie *The Day the Earth Stood Still* (2008), winged nanobots start eating everything on Earth.

HOW SMALL IS A NANOBOT?

The *nano* in nanobot comes from the word *nanometer*. A nanometer is one-billionth of a meter. A robot is considered a nanobot if it falls between 1 and 100 nanometers, but this is open to a bit of poetic license. If you had a nanobot that was 2,000 nanometers across, it would probably still fill the bill, although technically it is a microbot.

Here are the average diameters and lengths of different items to put this size range into perspective:

- A human hair is 100,000 nanometers in diameter.
- A human sperm cell is 2,500 nanometers in diameter (measured at the head), and 30,000 nanometers long.
- An *E. coli* cell is about 1,000 nanometers in diameter and 2,000 nanometers long.
- A virus particle, such as the flu virus, is about 120 nanometers in diameter.

- As of 2020, a single transistor is about 100 nanometers in diameter (with 7-nanometer technology, as presently used by Apple, considered state-of-the-art in 2020).

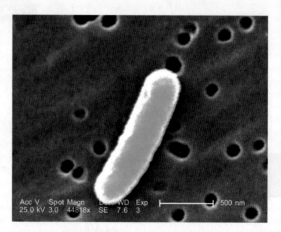

ABOVE A scanning electron micrograph of an *E. coli* cell.

Self-replication at something like a nano scale occurs in nature all the time. Bacteria are a great example. *E. coli* bacteria can duplicate themselves every 20 minutes under ideal conditions. To do so, an *E. coli* cell replicates its DNA strand, creates new cell wall (and all of the apparatus embedded in the wall), and copies all the structures inside the cell. For this to happen, the *E. coli* has to consume molecules containing the atoms of the elements it needs. These elements include hydrogen, oxygen, carbon, nitrogen, phosphorus, potassium, sulfur, magnesium, calcium, and iron. If the bacteria can find all ten of these elements in sufficient quantities and in accessible forms, it can reproduce.

Have humans ever created any example of self-replication? Nature obviously does it all the time, but what about human ingenuity? Could a car factory, for example, self-replicate? The answer at present is no, and there are two reasons why:

1 A car factory replicates cars, and it does this very well. A big car factory can produce hundreds of cars a day. But a car factory does not create new car factories. That would be a completely different beast.

2 A typical car factory today is not self-contained. It needs to take in steel from an iron mine and then a steel mill, plastic created from crude oil and then shaped by an injection molding company, paint from a paint manufacturer, glass from a glass manufacturer, tires from a tire maker, and so on. It is difficult to imagine a single, self-contained car factory that also creates all of the necessary materials it needs to create cars, much less to create another car factory, under one roof.

Right now, perhaps the closest thing that humans have to a self-replicator is a 3-D printer that prints another 3-D printer. This idea can work, to a degree, with today's technology. All the plastic pieces of a new 3-D printer can be manufactured using the printer, and the ability to print metal parts in the same way is in sight. But we will still get stuck when it comes time to make the silicon chips that control the 3-D printer. We would have to replicate the system that makes chips as well, and making silicon chips is a non-trivial, complex process. Finally, we need a machine to assemble the parts into a new printer, and a machine to make copies of this assembly machine.

Clearly, this is a complicated path.

What scientists are hoping to do is to leapfrog over these problems by working in a completely different domain. Nanobots that could work at the atomic level could in theory manipulate and arrange individual atoms. A nanobot that knows how to arrange certain atoms into a very specific configuration would be able to replicate itself.

Of course, this same atom-arranging nanobot could make anything else as well, as long as it had the atomic blueprint for it. Teams of millions of such nanobots could work together to create large objects. If scientists take humanity to this point, this could be beneficial as it drives the cost of any object toward zero. All that we need in order to produce anything is a batch of nanobots along with a supply of the needed atoms for the nanobots to manipulate, plus a blueprint.

RIGHT For a car factory to be self-replicating, it would need to manufacture the parts needed to assemble a car as well as the components that make up the structure of the factory itself.

PREVENTION

Is it possible for a self-replicating nanobot to wreak havoc in the science-fiction eat-a-whole-planet sense?

It is definitely possible to imagine creating a self-replicating nanobot, as described in the previous section. Also, we see self-replicating organisms in nature that are quite successful. Perhaps plankton are as close as nature has come to self-replication at a planet-wide scale. Plankton come in all shapes and sizes, and many of them are in the nano and micro realm. Plankton live in the world's oceans, so they have plenty of room to do their thing. Plankton get the chemicals they need for replication from the water in the ocean, and they use sunlight for energy. Even so, the world has not been overtaken by this prolific self-replicator. There are only so many necessary atoms mixed in with ocean water, and the sunlight only penetrates into the water so deep. Even plankton have their limits.

One way to prevent a nanobot gray goo from eating Planet Earth is to simply watch the nanobots hit an environmental constraint. Nanobots will run out of either chemicals or energy to replicate, just as everything in nature does. The nanobots might be annoying, but they will not eat the whole planet.

What gray goo might do, if programmed by a malevolent force, is cause a lot of problems. What if a malevolent nanobot lands on your skin, and what if your skin contains all the atomic elements the nanobot needs to replicate? What if the nanobot can derive energy from "eating" your skin or from your bloodstream just below the skin? And what if the nanobot starts replicating on your skin with the goal of eating you to create a cloud of new nanobots? This actually happens to a degree in the real world right now, and

we call it skin cancer. We might treat this nanobot the same way we treat skin cancer: excise it, freeze it, use chemicals to dissolve it, or poison it. We could also try to attack it with another beneficial nanobot or use a small electromagnetic pulse to disable it. There would be something humans could do to stop them.

But it is easy to imagine these flesh-eating nanobots jumping from humans to animals in the wild. And then what? Humans infected with nanobots can call 911 to receive treatment, while an animal in the wild cannot. If an evil genius were to also create a malevolent nanobot that eats trees, we'd definitely have a problem. There are trillions of trees on Earth, and we have witnessed biological organisms kill off lots of trees. Chestnut trees in the United States, for example, once numbered in the billions but are now extinct because of chestnut blight, a fungal infection. But even these tree-eating nanobots might be stopped, possibly by creating benevolent nanobots that eat the tree-eating nanobots.

What we can see with the gray goo scenario is a mixture of good news and bad news. Programmable matter would be fantastic if we can create it. But the idea of Earth being completely eaten by rogue nanobots would be tragic. However, it is unlikely to happen in reality because the nanobots will run into the same kinds of limits that nature's self-replicators do. On the other hand, self-replicating nanobots that eat all plant life on Earth would be devastating—and would require a massive and rapid response of some sort. The hope is that once humanity has the technology to create beneficial nanoscale self-replicators, it can also create the technology to stop malicious nanoscale self-replicators. We will likely be able to find a chink in their armor.

LEFT Like plankton, algae are also capable of rapid self-replication. When a high concentration of nutrients becomes available in the water and weather conditions and the salinity of the water are just right, they can reproduce fast enough to create an algal bloom known as a red tide.

CONCLUSION

As you come to the end of this book, you have seen twenty-five doomsday scenarios brought to life in vivid detail. Many of them portend massive destruction and upheaval with millions of lives lost. Some can seem almost inevitable. Unless things change significantly in the human psyche or in societal preparedness, we may watch helplessly as several of these scenarios unfold before our very eyes. Scenarios like chemical attacks, global warming, rainforest collapse, and the automation economy fall into this category.

You may find that this book makes you feel uncomfortable, possibly deeply despondent. Some of these scenarios truly are distressing, especially when they are collected together in one place and especially when we realize that humans are the direct cause of so much of this potential misery. Sometimes it is individual bad actors who cause big problems; other times it is masses of humans living their lives in intentional or unintentional ignorance of the destruction they are causing.

Here is the monumental question that this book poses: Can we as a species actually change our behavior to avoid every one of these doomsday scenarios? Can we take to heart and implement the types of preventive measures described in this book and save ourselves? Can we become proactive rather than reactive? Can we look all of these doomsday scenarios in the eye rationally, understand how to prevent them, and then take all the necessary steps?

The answer, with the current state of humanity as it stands today, is: probably not. We can see this with the COVID-19 pandemic. A pandemic of this severity had been predicted for decades, and we even had a few less disruptive outbreaks of similar diseases, such as SARS (severe acute respiratory syndrome) and MERS (Middle East respiratory syndrome), to practice on. Even so, many of the world's developed countries failed in their response when COVID-19 arrived, most notably the United States. Millions and millions of people contracted COVID-19, and so many people died unnecessarily. Our failure to prepare cost the global economy tens of trillions of dollars in medical costs, casualties, and lost productivity. Had we prepared properly and then performed expertly, most of this pain and death could have been completely avoided.

Or consider this: billions of people on Planet Earth today are blithely driving their gasoline-powered vehicles, as they have for more than a century. In the process, we release gigatons of carbon dioxide gas each month, and in the process we are actively destroying the planet in several doomsday ways. Add this to the gigatons of emissions from airplanes, trains, cargo ships, coal-fired power plants, and so many other things that burn fossil fuels. Every thoughtful person knows that this behavior is absolutely wrong, but it seems we are powerless to stop it.

Or think about the typical power grid in any developed nation. All of its important parts—its transmission lines, substations, and transformers—sit out in the open, unguarded, awaiting attack by terrorists or zealots. If individuals never bring the grid down, a coronal mass ejection could do it. If not a CME, an EMP attack might. With electricity so essential to technologically advanced societies, it is disheartening to think we are so vulnerable. Yet here we are.

Our behavior as a species must change if we want to prevent the doomsday scenarios described in this book from happening. For me, here is the thing that will reveal a new human condition and a completely new way of thinking for the human species: the day when all nations on Earth band together to launch a massive cooperative effort to end the use of fossil fuels planet-wide, combined with a massive effort to extract surplus carbon dioxide from the atmosphere in a short time frame. If humanity can tackle the end of fossil fuels, then perhaps we can also avoid pandemics with preparedness and technology, prevent nuclear holocaust with disarmament, confront the threat of asteroid strikes with a global space defense system, alleviate the terrible poverty that half of humanity is experiencing, and more. Humanity will be significantly better off if we can collaborate on this scale and solve all of the problems we face as a species.

Here's hoping that this book can help raise humanity's consciousness to a level where we can proactively address the many doomsday threats described in these pages.

ABOVE **All of life on Planet Earth hangs in the balance with many of these doomsday scenarios. Can humans step up and solve all the problems we face?**

ACKNOWLEDGMENTS

Have you ever wondered what goes into creating a book like this? Let's take a look . . . .

The Doomsday Book started with Kate Zimmermann, my editor for *The Engineering Book*. Kate introduced me to Sterling's new science editor, John Meils, and John and I started talking in December 2018. He asked me for some book ideas that I might be interested in writing, and I sent him a list of ten ideas and descriptions. After I submitted two formal book proposals in May 2019, we settled on *The Doomsday Book* as my next project..

The first draft of *The Doomsday Book* was written during the summer and fall of 2019. My family (six of us and our dog) happened to be between houses and living in a tiny apartment in Cary, North Carolina. Much of this book's manuscript was written in the kitchen of this little apartment, usually at 4 a.m. and on weekends. We moved into a new house in the middle of the first draft, which added bonus levels of fun to the process.

I am extremely grateful to John for believing in this project, creating the four-part chapter template used throughout this book, and encouraging me so I got the manuscript submitted on time in spite of the apartment and the move.

On the day that I delivered the final pieces of the first draft, John announced that he was moving to another company. I met Elysia Liang, who became the editor for this book after John's exit. A good editor is a combination of project manager, schedule wrangler, coordinator of many people, and skillful negotiator, as well as the person who edits the manuscript. In the case of this book, the editing process involved reviewing three passes (a first edit, second edit, and copyedit) and layouts, and consisted of an intense dialogue of questions, feedback, cuts, and rewrites. And let's not forget that COVID-19 struck right in the middle of this process and that Elysia and Sterling are based in New York City, the pandemic's first major epicenter in the United States. Elysia had to manage all the disruptions and weirdness brought on by a doomsday scenario that jumped right off the pages and came to life in real time as we were producing this book. I cannot possibly thank Elysia enough for everything she has brought to this project.

It is a big team effort to produce a book like this. I am incredibly grateful to everyone who made this book possible, including Elizabeth Lindy for the beautiful cover design, Christine Heun for the incredible layout for the interior of this book, Linda Liang for finding the artwork, and Scott Amerman and Michael Cea for managing the copyediting, proofreading, and indexing.

The book went to press in July 2020 and was produced in Spain, ready for its launch in October 2020.

SELECTED BIBLIOGRAPHY

For the full bibliography, please head to DoomesdayNow.com. This website is a companion for this book and contains links to multimedia content to enhance your experience. In addition, you'll find a blog that keeps you up to date on the latest news related to the doomsday scenarios in this book.

INTRODUCTION

Boxer, Sarah. "EYEWITNESSES: One Camera, Then Thousands, Indelibly Etching a Day of Loss." *New York Times*, September 11, 2002. https://www.nytimes.com/2002/09/11/us/eyewitnesses-one-camera-then-thousands-indelibly-etching-a-day-of-loss.html.

Cachero, Paulina. "US Taxpayers Have Reportedly Paid an Average of $8,000 Each and Over $2 trillion Total for the Iraq War Alone." Business Insider, February 6, 2020. https://www.businessinsider.com/us-taxpayers-spent-8000-each-2-trillion-iraq-war-study-2020-2.

Crawford, Neta C. "Human Cost of the Post-9/11 Wars: Lethality and the Need for Transparency," November 2018. Costs of War. https://watson.brown.edu/costsofwar/files/cow/imce/papers/2018/Human%20Costs%2C%20Nov%208%202018%20CoW.pdf.

de Castella, Tom. "Who, What, Why: How Are Cockpit Doors Locked?" *BBC*, March 26, 2015. https://www.bbc.com/news/blogs-magazine-monitor-32070528.

Grossman, David. "Destroyed on 9/11, a New York Subway Station Reopens 17 Years Later." *Popular Mechanics*, September 10, 2018. https://www.popularmechanics.com/technology/infrastructure/a23064589/new-york-wtc-cortlandt-subway-september-11/.

Hirschkorn, Phil, and Peter Bergen. "Rare Photos Reveal Osama bin Laden 's Afghan Hideout." CNN, March 18, 2015. https://www.cnn.com/2015/03/11/world/osama-bin-laden-hideout-photos/index.html.

"How Osama bin Laden and Al Qaeda Planned 9/11." History.com. Accessed April 24, 2020. Video, 3:07. https://www.history.com/topics/21st-century/birth-of-al-qaeda-video.

Kiger, Patrick J. "Pearl Harbor: Photos and Facts from the Infamous WWII Attack." History.com. Last modified December 7, 2018. https://www.history.com/news/pearl-harbor-facts-wwii-attack.

Locker, Ryan. "10 Things You May Have Forgotten about 9/11." *USA Today*. Last modified September 11, 2017. https://www.usatoday.com/story/news/politics/2016/09/10/10-things-you-may-have-forgotten-911/90007376/.

SPLITTING THE UNITED STATES IN HALF

Ashe, Ari, and Hugh R. Morley. "US East Coast Ports Investing to Capture More Intermodal Cargo." JOC.com, January 27, 2020. https://www.joc.com/rail-intermodal/intermodal-shipping/us-east-coast-ports-investing-capture-more-intermodal-cargo_20200127.html.

Bush, Mike. "'Flawed' Dam May Threaten St. Louis Area." *KSDK-TV*, April 28, 2017. https://www.ksdk.com/article/news/local/storytellers/flawed-dam-may-threaten-st-louis-area/63-434663309.

Frey, Thomas. "The Fort Peck Incident." Futurist Speaker, July 10, 1998. https://futuristspeaker.com/future-scenarios/the-fort-peck-incident/.

Largest Dams. "St. Francis Dam Disaster." YouTube video, 3:21. March 24, 2018. https://www.youtube.com/watch?v=Tp7sWZmk7gI.

Largest Dams. "Teton Dam Disaster." YouTube video, 2:47. March 24, 2018. https://www.youtube.com/watch?v=nQ0MyBg5h_A.

Mwdh2o. "A Timeline of Oroville Events–2017." YouTube video, 7:26. April 24, 2017. https://www.youtube.com/watch?v=NjbbW37qzak.

National Institute of Standards and Technology. "NIST Video: Why the Building (WTC7) Fell." YouTube video, 3:39. January 28, 2009. https://www.youtube.com/watch?v=PK_iBYSqEsc.

Smithsonian Channel. "One of the Worst Man-Made Disasters in History." YouTube video, 4:04. November 3, 2017. https://www.youtube.com/watch?v=lkGnnc8Ezlk.

DRONE STRIKES AND SWARMS

Asia News. "Drone Taxi Dubai." YouTube video, 1:39. February 14, 2017. https://www.youtube.com/watch?v=5Rfe4BFiVNA.

Berger, Sarah. "Mexico Drug Trafficking: Drone Carries 28 Pounds of Heroin Across Border to US." *International Business Times*, August 13, 2015. https://www.ibtimes.com/mexico-drug-trafficking-drone-carries-28-pounds-heroin-across-border-us-2051941.

Bloomberg Markets and Finance. "First Look at Damage From Drone Attack on Saudi Aramco Facility." YouTube video, 1:24. September 20, 2019. https://www.youtube.com/watch?v=wjKGtXsDwqc.

"China Launches Record-Breaking UAV Swarm." *Pakistan Defense*, June 25, 2017, https://defence.pk/pdf/threads/china-launches-record-breaking-uav-swarm.503463.

CNN. "Inside the Drone Attack That Nearly Killed a World Leader." YouTube video, 5:38. March 14, 2019. https://www.youtube.com/watch?v=EPgbKRgYwpc.

Conger, Kate. "How Consumer Drones Wind Up in the Hands of ISIS Fighters." *Tech Crunch*, October 13, 2016. https://techcrunch.com/2016/10/13/how-consumer-drones-wind-up-in-the-hands-of-isis-fighters/.

"Defense against Drones." Franhofer. Accessed April 22, 2020. https://www.fraunhofer.de/en/research/current-research/defense-against-drones.html.

Defense Updates. "IRON BEAM: ISRAEL'S LASER AIR DEFENSE SYSTEM: TOP 5 FACTS." YouTube video, 1:39. June 29, 2015. https://www.youtube.com/watch?v=OEpZB2OkRh0.

DiGiulian, Tony. "20 mm Phalanx Close-in Weapon System (CIWS)." NavWeaps. Accessed April 22, 2020. http://www.navweaps.com/Weapons/WNUS_Phalanx.php.

Flynt, Joseph. "How Much Weight Can a Drone Carry?" 3D Insider, November 29, 2017. https://3dinsider.com/drone-payload/.

Fogerlie, Garrett. "Boeing's CHAMP Missile Demo–Directed High Frequency Microwave Disables Electronics (EMP)." YouTube video, 2:08. February 24, 2013. https://www.youtube.com/watch?v=Lh1rgy25XhU.

"Iran Presents Its Suicide Drones." The Middle East Media Research Institute, April 10, 2015. https://www.memri.org/reports/iran-presents-its-suicide-drones.

Lin, Jeffrey, and P. W. Singer. "China's New Microwave Weapon Can Disable Missiles and Paralyze Tanks." *Popular Science*, January 26, 2017. https://www.popsci.com/china-microwave-weapon-electronic-warfare/.

Marcolini, Barbara. "Video Shows Drone Attack on Maduro in Venezuela." *New York Times*, August 5, 2018. Video, 1:23. https://www.nytimes.com/video/world/americas/100000006040957/venezuela-maduro-drone-attack.html.

McNabb, Miriam. "Drone at Prisons: The Results of This 9-Month Study Show Exactly What We Need to Counter Drone Tech." Drone Life, March 15, 2019. https://dronelife.com/2019/03/15/drones-at-prisons-the-results-of-this-9-month-study-show-exactly-why-we-need-counter-drone-technology/.

Murdock, Jason. "'Drone Swarm' Used by Criminals to Disrupt an FBI Hostage Rescue Operation." *Newsweek*, May 14, 2018. https://www.newsweek.com/drone-swarm-used-criminals-disrupt-fbi-hostage-rescue-operation-910431.

O'Brien, Matt. "Army Looks for a Few Good Robots, Sparks Industry Battle." *Army Times*, December 30, 2018. https://www.armytimes.com/news/your-army/2018/12/30/army-looks-for-a-few-good-robots-sparks-industry-battle/.

"Photos Show Precision of Attack on Saudi Oil Plant." *Daily Sabah*, September 16, 2019. https://www.dailysabah.com/mideast/2019/09/16/photos-show-precision-of-attack-on-saudi-oil-plant.

Singer, Peter W. "Defending against Drones." Brookings Institute, February 28, 2010. https://www.brookings.edu/articles/defending-against-drones/.

South China Morning Post. "Saudi Arabia's Oil Output Decimated by Drone Attack." YouTube video, 2:12. September 16, 2019. https://www.youtube.com/watch?v=ryIZPLsYjCE.

"Tomahawk Long-Range Cruise Missile." Navel Technology. Accessed April 22, 2020. https://www.naval-technology.com/projects/tomahawk-long-range-cruise-missile/.

Trevithick, Joseph. "These Pods Could Let Drones Carry Throngs of Small Smart Bombs And It's Just The Start." The Drive, October 8, 2018. https://www.thedrive.com/the-war-zone/24129/these-pods-could-let-drones-carry-throngs-of-small-smart-bombs-and-its-just-the-start.

Waterman, Shaun. "Directed Energy Weapons Move Closer to Prime Time." *Air Force Magazine*, October 29, 2019. https://www.airforcemag.com/directed-energy-weapons-move-closer-to-prime-time/.

Watson, Ben. "Against the Drones." Defense One. Accessed April 22, 2020. https://www.defenseone.com/feature/against-the-drones/.

"What Defense Is There against Off-the-Shelf Drone Weapons?" DW. Accessed April 22, 2020. https://www.dw.com/en/what-defense-is-there-against-off-the-shelf-drone-weapons/a-44970742.

NUCLEAR BOMBS

"Aegis: The Shield of the Fleet." Lockheed. Accessed May 19, 2020. https://www.lockheedmartin.com/en-us/products/aegis-combat-system.html.

"The B83 (Mk-83) Bomb—High Yield Strategic Thermonuclear Bomb." Nuclear Weapon Archive, November 11, 1997. http://nuclearweaponarchive.org/Usa/Weapons/B83.html.

"Critical Mass." Atomic Archive. http://www.atomicarchive.com/Fission/Fission3.shtml.

Davenport, Kelly. "Nuclear Weapons: Who Has What at a Glance." Arms Control Association, July 2019. https://www.armscontrol.org/factsheets/Nuclearweaponswhohaswhat.

Derouin, Sarah. "Nuclear Winter May Bring a Decade of Destruction." Eos, September 27, 2019. https://eos.org/articles/nuclear-winter-may-bring-a-decade-of-destruction.

Federal Emergency Management Agency. *Building Design for Homeland Security: Unit VI, Explosive Blast.* https://www.fema.gov/pdf/plan/prevent/rms/155/e155_unit_vi.pdf.

Fitzgerald, Sunny. "Nuclear Missile Aftermath: How People in Hawaii Reacted When They Thought They Had 15 Minutes Left." *Mic*, January 23, 2018. https://www.mic.com/articles/187537/nuclear-missile-aftermath-how-people-in-hawaii-reacted-when-they-thought-they-had-15-minutes-left.

Glaser, Alexander. "Effects of Nuclear Weapons." Princeton University, February 12, 2007. https://www.princeton.edu/~aglaser/lecture2007_weaponeffects.pdf.

Gordon, Elizabeth. "The Manhattan Project—Critical Mass and Bomb Construction." LibreTexts, February 26, 2020. https://chem.libretexts.org/Courses/Furman_University/CHM101%3A_Chemistry_and_Global_Awareness_(Gordon)/06%3A_Nuclear_Weapons-_Fission_and_Fusion/6.4%3A_The_Manhattan_Project_-_Critical_Mass_and_Bomb_Construction.

Jabr, Ferris. "This Is What a Nuclear Bomb Looks Like." *New York Magazine*, June 11, 2018. http://nymag.com/intelligencer/2018/06/what-a-nuclear-attack-in-new-york-would-look-like.html.

Jervis, Robert. "The Dustbin of History: Mutual Assured Destruction." *Foreign Policy*, November 9, 2009. https://foreignpolicy.com/2009/11/09/the-dustbin-of-history-mutual-assured-destruction/.

"Largest Cities in the United States by Population." Ballotpedia, 2013. https://ballotpedia.org/Largest_cities_in_the_United_States_by_population.

"Little Boy and Fat Man." Atomic Heritage Foundation, July 23, 2014. https://www.atomicheritage.org/history/little-boy-and-fat-man.

Nagourney, Adam, David E. Sanger, and Johanna Barr. "Hawaii Panics After Alert about Incoming Missile Is Sent in Error." *New York Times*. January 13, 2018. https://www.nytimes.com/2018/01/13/us/hawaii-missile.html.

"North Korea Nuclear Weapons Threat." Nuclear Threat Initiative, August 2019. https://www.nti.org/learn/countries/north-korea/.

"Nuclear Weapons Primer." Wisconsin Project on Nuclear Arms Control. Accessed May 19, 2020. https://www.wisconsinproject.org/nuclear-weapons/.

"Nuclear Winter." Atomic Archive. Accessed May 19, 2020. https://www.atomicarchive.com/Effects/effects23.shtml.

Perez, Maria. "Chilling Video: Child Placed in Storm Drain During Hawaii Missile Scare." *Newsweek*, January 15, 2018. https://www.newsweek.com/hawaii-ballistic-missile-north-korea-us-781535.

"Plutonium-239 Formation." Radioactivity.eu.com. Accessed May 19, 2020. http://www.radioactivity.eu.com/site/pages/Plutonium_239_Formation.htm.

"The 'Shadow' of a Hiroshima Victim, Etched into Stone Steps, Is All That Remains After 1945 Atomic Blast." Open Culture, March 16, 2016. http://www.openculture.com/2016/03/the-shadow-of-a-hiroshima-victim-etched-into-stone-steps-is-all-That-remains-after-1945-atomic-blast.html.

"THAAD: Integrated Air and Missile Defense with Proven Hit-to-Kill Technology." Lockheed. Accessed May 19, 2020. https://www.lockheedmartin.com/en-us/products/thaad.html.

Ulmer, Dana S. "Uranium—Where Is It Found?" New Mexico Bureau of Geology and Mineral Resources. Accessed May 19, 2020. https://geoinfo.nmt.edu/resources/uranium/where.html.

"United States Nuclear Capabilities." Nuclear Threat Initiative, July 2017. https://www.nti.org/learn/countries/united-states/nuclear/.

"Uranium-235 Chemical Isotope." Encyclopedia Britannica. https://www.britannica.com/science/uranium-235.

U.S. Department of Defense Missile Defense Agency. Accessed May 19, 2020. https://www.mda.mil/.

Wellerstein, Alex. "Nukemap." The Nuclear Secrecy Blog. Accessed May 19, 2020. https://nuclearsecrecy.com/nukemap/.

"'The Whole State Was Terrified': How Hawaii Reacted to False Missile Alert." *BBC*, January 13, 2018. https://www.bbc.com/news/world-us-canada-42675666.

Wisti, Erin. "11 Haunting Photos of Shadows Permanently Burned into the Ground by the Hiroshima Nuclear Blast." Ranker, April 22, 2020. https://www.ranker.com/list/photos-of-shadows-burned-into-hiroshima/erin-wisti.

ANTARCTICA COLLAPSE

"Antarctica Weather and Climate." Cool Antarctica. Accessed May 1, 2020. https://www.coolantarctica.com/Antarctica%20fact%20file/antarctica%20environment/climate_weather.php.

Barnard, Anne. "The $119 Billion Sea Wall That Could Defend New York . . . or Not." *New York Times*, January 17, 2020. https://www.nytimes.com/2020/01/17/nyregion/sea-wall-nyc.html.

Beeler, Carolyn. "If Thwaites Glacier Collapses, It Would Change the Global Coastlines Forever." *The World*, July 1, 2019. https://interactive.pri.org/2019/05/antarctica/thwaites-glacier-collapse.html.

Bergstrom, Dana M., Andrew Klekociuk, Diana King, and Sharon Robinson. "Antarctica Has Just Had Its First Recorded Heatwave." *Newsweek*, March 31, 2020. https://www.newsweek.com/heatwave-antarctica-recorded-first-time-1495206.

Chen, Angela. "Use These Tools to Help Visualize the Horror of Rising Sea Levels." *The Verge*, February 17, 2019. https://www.theverge.com/2019/2/17/18223808/climate-change-sea-level-rising-data-visualization-environment.

Docquier, David. "What's Up on Thwaites Glacier?" EGU Blogs, March 13, 2020. https://blogs.egu.eu/divisions/cr/2020/03/13/whats-up-on-thwaites-glacier/.

Flamm, Matthew. "The End of the Hudson as We Know It?" *Crain's New York Business*, May 19, 2019. https://www.crainsnewyork.com/features/end-hudson-we-know-it.

Gertner, Jon. "The Race to Understand Antarctica's Most Terrifying Glacier." *Wired*, December 10, 2018. https://www.wired.com/story/antarctica-thwaites-glacier-breaking-point/.

Harvey, Chelsea. "Antarctic Melt Rate Has Tripled in the Last 25 Years." *Scientific American*, June 14, 2018. https://www.scientificamerican.com/article/antarctic-melt-rate-has-tripled-in-the-last-25-years/.

Hickey, Hannah. "West Antarctic Ice Sheet Collapse Is Under Way." *University of Washington News*, May 12, 2014. https://www.washington.edu/news/2014/05/12/west-antarctic-ice-sheet-collapse-is-under-way/.

Irfan, Umair. "Antarctica Broke Two Temperature Records in a Week." *Vox*, February 13, 2020. https://www.vox.com/2020/2/7/21128389/antarctica-record-heat-temperature-climate-change-glacier-ice-melt.

Lewis, Sophie. "Scientists Alarmed to Discover Warm Water at 'Vital Point' beneath Antarctica's 'Doomsday Glacier.'" *CBS News*, February 1, 2020. https://www.cbsnews.com/news/climate-change-thwaites-melting-scientists-warm-water-antarctica-doomsday-glacier/.

McKie, Robin. "Scientists Discover 91 Volcanoes below Antarctic Ice Sheet." *The Guardian*, August 12, 2017. https://www.theguardian.com/world/2017/aug/12/scientists-discover-91-volcanos-antarctica.

Niewenhuis, Lucas. "Rising Sea Levels to Turn Shanghai into a Swamp—Study." Supchina, October 31, 2019. https://supchina.com/2019/10/31/rising-sea-levels-to-turn-shanghai-into-a-swamp-study/.

Rice, Doyle. "Warm Water Discovered beneath Antarctica's 'Doomsday' Glacier,' Scientists Say." *USA Today*, February 3, 2020. https://www.usatoday.com/story/news/nation/2020/01/31/thwaites-glacier-warm-water-discovered-below-doomsday-glacier/4621338002/.

Smith, Kiona N. "40-Year-Old Radar Data Shows Antarctica's Thwaites Glacier Is Melting Faster Than We Realized." *Forbes*, September 20, 2019. https://www.forbes.com/sites/kionasmith/2019/09/20/40-year-old-radar-data-shows-antarcticas-thwaites-glacier-is-melting-faster-than-we-realized/#4512939a4548.

"What Is Iceberg Calving, and Why Does It Matter?" CPOM. Accessed May 1, 2020. https://cpom.org.uk/what-is-iceberg-calving-and-why-does-it-matter/.

Woodward, Aylin. "Ice Sheets Are Melting Far Faster Than We Thought—in a Worst-case Climate Breakdown, Coastal Cities Like New York and Shanghai Would Be Swamped." Business Insider, May 21, 2019. https://www.businessinsider.com/sea-level-rise-could-swamp-new-york-shanghai-2019-5.

EMP ATTACK

Blair, Christopher W., Casey Mahoney, Shira E. Pindyck, and Joshua A. Schwartz. "Trump Issued an Executive Order to Prepare for an EMP Attack. What Is It, and Should You Worry?" *Washington Post*, March 29, 2019. https://www.washingtonpost.com/politics/2019/03/29/trump-issued-an-executive-order-prepare-an-emp-attack-what-is-it-should-you-worry/.

Butt, Yousaf M. "The EMP Threat: Fact, Fiction, and Response (part 1)." *The Space Review*, January 25, 2010. http://www.thespacereview.com/article/1549/1.

Commission to Assess the Threat to the United States from Electromagnetic Pulse (EMP) Attack. "Home." Accessed April 22, 2020. http://www.empcommission.org/.

Commission to Assess the Threat to the United States from Electromagnetic Pulse (EMP) Attack. "Report of the Commission to Assess the Threat to the United States from Electromagnetic Pulse (EMP) Attack." 2008. http://www.empcommission.org/docs/A2473-EMP_Commission-7MB.pdf.

Congressional Research Service. *High Altitude Electromagnetic Pulse (HEMP) and High Power Microwave (HPM) Devices: Threat Assessments* by Clay Wilson. RL32544. 2008. https://www.wired.com/images_blogs/dangerroom/files/Ebomb.pdf.

Emanuelson, Jeremy. "E1, E2 and E3." Futerscience, LLC. Accessed April 22, 2020. http://www.futurescience.com/emp/E1-E2-E3.html.

Federal Energy Regulatory Commission. *Electromagnetic Pulse: Effects on the U.S. Power Grid.* https://www.ferc.gov/industries/electric/indus-act/reliability/cybersecurity/ferc_executive_summary.pdf.

Gault, Matthew. "Would 90 Percent of Americans Really Die from an EMP Attack? Some Think So." *The National Interest*, January 10, 2018. https://nationalinterest.org/blog/the-buzz/would-90-percent-americans-really-die-emp-attack-some-think-24005.

Golijan, Rosa. "This EMP Cannon Stops Cars Almost Instantly." *Gizmodo*, January 21, 2010. https://gizmodo.com/this-emp-cannon-stops-cars-almost-instantly-5454295.

Hambling, David. "What Is an EMP, and Could North Korea Really Use One Against the U.S.?" *Popular Mechanics*, September 28, 2017. https://www.popularmechanics.com/military/weapons/news/a28425/emp-north-korea/.

Maloney, Dan. "How to Test a B-52 against EMP: Project Atlas-I." Hackaday, modified March 9, 2018. https://hackaday.com/2018/03/09/how-to-test-a-b-52-against-emp-project-atlas-i/.

Mosher, Dave. "Nuclear Bombs Trigger a Strange Effect That Can Fry Your Electronics—Here's How It Works" Business Insider, June 7, 2017. https://www.businessinsider.com/nukes-electromagnetic-pulse-electronics-2017-5.

Plait, Phil. "The 50th Anniversary of Starfish Prime: The Nuke That Shook the World." *Discover Magazine*, July 9, 2012. https://www.discovermagazine.com/the-sciences/the-50th-anniversary-of-starfish-prime-the-nuke-That-shook-the-world.

"States Work to Protect the Electric Grid from Catastrophic Disturbances." EMPact America. Accessed May 1, 2020. https://empactamerica.org/states-work-protect-electric-grid-catastrophic-disturbances/.

Subcommittee on the National Subcommittee on the Interior House Committee on Oversight and Government Reform. *The EMP Threat: The State of Preparedness Against the Threat of a Electromagnetic Pulse (EMP) Event* by Vincent Pry. 2015. https://republicans-oversight.house.gov/wp-content/uploads/2015/05/Pry-Statement-5-13-EMP.pdf.

Subcommittee on Oversight and Management Efficiency of the Committee on Homeland Security House of Representatives. *Empty Threat or Serious Danger: Assessing North Korea's Risk to the Homeland*. 2017. https://docs.house.gov/meetings/HM/HM09/20171012/106467/HHRG-115-HM09-Transcript-20171012.pdf.

U.S. Department of Homeland Security. National Coordinating Center for Communications (NCC) and National Cybersecurity & Communications Integration Center (NCCIC). *Electromagnetic Pulse (EMP) Protection and Restoration Guidelines for Equipment and Facilities*. 2016. https://info.publicintelligence.net/DHS-FacilitiesGuidelinesEMP.pdf.

Voute, Christopher. "The Theory Behind EMP Generators." Wonder How To, February 5, 2012. https://fear-of-lightning.wonderhowto.com/how-to/making-electromagnetic-weapons-theory-behind-emp-generators-0133121/.

Wilson, Jim. "E-Bomb." *Popular Mechanics*, September 2001, 51–53.

RUNAWAY GLOBAL WARMING

Allen, C. D. "Climate-Induced Forest Dieback: An Escalating Global Phenomenon?" Food and Agriculture Organization of the United Nations. Accessed May 1, 2020. http://www.fao.org/3/i0670e10.htm.

Ayres, Robert. "The Ocean Cannot Absorb Much More CO2." Insead, October 19, 2016. https://knowledge.insead.edu/blog/insead-blog/the-ocean-cannot-absorb-much-more-co2-4990.

Berwyn, Bob. "Heat Waves Creeping Toward a Deadly Heat-Humidity Threshold." *Inside Climate News*, August 3, 2017. https://insideclimatenews.org/news/02082017/heatwaves-deadly-heat-humidity-wet-bulb-human-survivability-threshold.

Casben, Liv. "Summers Are Now Twice as Long as Winters in All Australian Capital Cities, Report Finds." *ABC News*, March 1, 2020. https://www.abc.net.au/news/2020-03-02/australian-summers-getting-longer-winters-shorter/12013978.

Childs, Samuel. "Destructive 2018 Hail Season a Sign of Things to Come." *The Conversation*, September 20, 2018. https://theconversation.com/destructive-2018-hail-season-a-sign-of-things-to-come-102879.

Cho, Renee. "How We Know Today's Climate Change Is Not Natural." State of the Planet, April 4, 2017. https://blogs.ei.columbia.edu/2017/04/04/how-we-know-climate-change-is-not-natural/.

Chow, Lorraine. "10 Worst-Case Climate Predictions If Global Temperatures Rise Above 1.5 Degrees Celsius." Common Dreams, January 2, 2019. https://www.commondreams.org/views/2019/01/02/10-worst-case-climate-predictions-if-global-temperatures-rise-above-15-degrees.

Dunne, Daisy. "Explainer: Six Ideas to Limit Global Warming with Solar Geoengineering." CarbonBrief, September 5, 2018. https://www.carbonbrief.org/explainer-six-ideas-to-limit-global-warming-with-solar-geoengineering.

Fujita, Rod. "5 Ways Climate Change Is Affecting Our Oceans." Environmental Defense Fund, October 8, 2013. https://www.edf.org/blog/2013/10/08/5-ways-climate-change-affecting-our-oceans.

Gorvett, Zaria. "How a Giant Space Umbrella Could Stop Global Warming." *BBC*, April 26, 2016. https://www.bbc.com/future/article/20160425-how-a-giant-space-umbrella-could-stop-global-warming.

Hickey, Hannah. "Could Spraying Particles into Marine Clouds Help Cool the Planet?" University of Washington, July 25, 2017. https://www.washington.edu/news/2017/07/25/could-spraying-particles-into-marine-clouds-help-cool-the-planet/.

Irfan, Umair. "Greenland's Ice Sheet Is Melting at Its Fastest Rate in Centuries." *Vox*, December 10, 2018. https://www.vox.com/energy-and-environment/2018/12/8/18129132/greenland-ice-melt-sea-level-climate-change.

Lewis, Sophie. "Planting a Trillion Trees Could Be the 'Most Effective Solution' to Climate Change, Study Says." *CBS News*, July 8, 2019. https://www.cbsnews.com/news/planting-a-trillion-trees-could-be-the-most-effective-solution-to-climate-change/.

Lizotte, Melissa. "County Farmers Experiencing Crop Challenges due to Increased Rainfall, Dry Weather." *The County*, February 24, 2020. https://thecounty.me/2020/02/24/news/county-farmers-experiencing-crop-challenges-due-to-increased-rainfall-dry-weather/.

MacDonald, Fiona. "This Unsettling Animation Shows What Earth Would Look Like If All the Ice Melted." *Science Alert*, June 19, 2018. https://www.sciencealert.com/this-animation-shows-what-earth-would-look-like-if-all-the-ice-melted.

McNeill, Leila. "This Lady Scientist Defined the Greenhouse Effect but Didn't Get the Credit, Because Sexism." *Smithsonian Magazine*, December 5, 2016. https://www.smithsonianmag.com/science-nature/lady-scientist-helped-revolutionize-climate-science-didnt-get-credit-180961291/.

McSweeney, Robert. "Explainer: Nine 'Tipping Points' That Could Be Triggered by Climate Change." CarbonBrief, February 10, 2020. https://www.carbonbrief.org/explainer-nine-tipping-points-That-could-be-triggered-by-climate-change.

"Methane and Frozen Ground." NISDC. Accessed May 9, 2019. https://nsidc.org/cryosphere/frozenground/methane.html.

Meyer, Robinson. "Are We Living through Climate Change's Worst-Case Scenario?" *The Atlantic*, January 15, 2019. https://www.theatlantic.com/science/archive/2019/01/rcp-85-the-climate-change-disaster-scenario/579700/.

Mooney, Chris, and Brady Dennis. "Staggering New Data Show Greenland's Ice Is Melting 6 Times Faster Than in the 1980s." *Science Alert*, April 23, 2019. https://www.sciencealert.com/greenland-s-accelerating-ice-loss-is-worrisome-to-scientists.

Mulligan, James, Gretchen Ellison, and Kelly Levin. "6 Ways to Remove Carbon Pollution from the Sky." Worlds Resources Institute, September 10, 2018. https://www.wri.org/blog/2018/09/6-ways-remove-carbon-pollution-sky.

Pearce, Fred. "As Climate Change Worsens, A Cascade of Tipping Points Looms." YaleEnvironment360, December 6, 2019. https://e360.yale.edu/features/as-climate-changes-worsens-a-cascade-of-tipping-points-looms.

Pearce, Fred. "Geoengineer the Planet? More Scientists Now Say It Must Be an Option." YaleEnvironment360, May 29, 2019. https://e360.yale.edu/features/geoengineer-the-planet-more-scientists-now-say-it-must-be-an-option.

Rohde, Robert. "Global Temperature Report for 2018." Berkeley Earth, January 24, 2019. http://berkeleyearth.org/2018-temperatures/.

Rodriguez Mega, Emiliano. "Clouds' Cooling Effect Could Vanish in a Warmer World." *Nature*, February 26, 2019. https://www.nature.com/articles/d41586-019-00685-x.

Rotman, David. "A Cheap and Easy Plan to Stop Global Warming." *MIT Technology Review*, February 8, 2013. https://www.technologyreview.com/2013/02/08/84239/a-cheap-and-easy-plan-to-stop-global-warming/.

Samuel, Sigal. "The Case for Spraying (Just Enough) Chemicals into the Sky to Fight Climate Change." *Vox*, March 13, 2019. https://www.vox.com/future-perfect/2019/3/13/18263953/geoengineering-study-solar-dose-climate-global-warming.

Tans, Pieter, and Kirk Thoning. "How We Measure Background CO2 Levels on Mauna Loa." NOAA, March 2018. https://www.esrl.noaa.gov/gmd/ccgg/about/co2_measurements.html.

Teitel, Amy Shira. "The Moist Greenhouse Effect Could Be the One That Destroys Our Atmosphere." *Vice*, July 6, 2013. https://www.vice.com/en_us/article/gvva5m/the-moist-greenhouse-effect-could-be-the-one-That-destroys-our-atmosphere.

Wallace-Wells, David. "The Uninhabitable Earth." *New York Magazine*, July 2017. https://nymag.com/intelligencer/2017/07/climate-change-earth-too-hot-for-humans.html.

Wallace-Wells, David. "UN Says Climate Genocide Is Coming. It's Actually Worse Than That." *New York Magazine*, October 10, 2018. https://nymag.com/intelligencer/2018/10/un-says-climate-genocide-coming-but-its-worse-than-That.html.

PANDEMICS AND BIOLOGICAL ATTACKS

Acuna-Soto, Rodolfo, Stahle, David, Cleaveland, Malcolm, and Therrell, Matthew. "Megadrought and Megadeath in 16th Century Mexico." *Emergent Infectious Diseases* 8, no. 4 (April 2002): 360–362. https://www.ncbi.nlm.nih.gov/pmc/articles/PMC2730237/.

Anderson, Andrea. "Smallpox Evolution, History Explored With Genome Sequence from Ancient Remains." Genome Web, December 8, 2016. https://www.genomeweb.com/sequencing/smallpox-evolution-history-explored-genome-sequence-ancient-remains.

"Austria Has 90% Drop in Coronavirus Cases After Requiring People to Wear Face Masks." *The Science Times*, April 21, 2020. https://www.sciencetimes.com/articles/25410/20200421/austria-90-drop-coronavirus-cases-requiring-people-wear-face-masks.htm.

Balz, Dan. "Once Again, Government Is Caught Unprepared." *Stamford Advocate*, April 4, 2020. https://www.stamfordadvocate.com/news/article/Once-again-government-is-caught-unprepared-15179581.php.

Bhaskaran, Prathapan. "Coronavirus Reshapes Global Economy: Japan To Fund Companies To Shift Production Out of China." *International Business Times*, April 10, 2020. https://www.ibtimes.com/coronavirus-reshapes-global-economy-japan-fund-companies-shift-production-out-china-2956336.

Biesecker, Michael. "US 'Wasted' Months before Preparing for Coronavirus Pandemic." *Associated Press*, April 6, 2020. https://apnews.com/090600c299a8cf07f5b44d92534856bc.

"China Coronavirus: Lockdown Measures Rise across Hubei Province." *BBC*, January 23, 2020. https://www.bbc.com/news/world-asia-china-51217455.

"The Coronavirus Explained & What You Should Do." Youtube video, 8:34. Posted by "Kurzgesagt–In a Nutshell," March 19, 2020. https://youtu.be/BtN-goy9VOY.

"Covid-19 to Wipe Out Equivalent of 195m Jobs, Says UN Agency." *The Guardian*, April 7, 2020. https://www.theguardian.com/world/2020/apr/07/covid-19-expected-to-to-wipe-out-67-of-worlds-working-hours.

Dunn, Will. "How America Built the Best Pandemic Response System in History—and Threw It Away." *New Statesman*, March 24, 2020. https://www.newstatesman.com/world/north-america/2020/03/america-pandemic-response-swine-flu-avian.

Dupuis, Adam. "HIV Arrived In U.S. Long Before 'Patient Zero' Did. History of HIV/AIDS n America Rewritten." *Instinct*, October 26, 2016. https://instinctmagazine.com/hiv-arrived-in-u-s-long-before-patient-zero-did-history-of-hiv-aids-in-america-rewritten/.

Fan, Ying, Sumana Sanyal, and Roberto Bruzzone. "Breaking Bad: How Viruses Subvert the Cell Cycle." *Frontiers in Cellular and Infection Microbiology*, November 19, 2018. https://www.frontiersin.org/articles/10.3389/fcimb.2018.00396/full.

Flight, Colette. "Smallpox: Eradicating the Scourge." *BBC*, February 17, 2011. https://www.bbc.co.uk/history/british/empire_seapower/smallpox_01.shtml.

France-Presse, Agence. "500 Years Later, Scientists Discover What Probably Killed the Aztecs." *The Guardian*, January 15, 2018. https://www.theguardian.com/world/2018/jan/16/mexico-500-years-later-scientists-discover-what-killed-the-aztecs.

Gill, Jr., Harold B. "Colonial Germ Warfare." *Colonial Williamsburg Journal*, 2004. Accessed October 28, 2019. https://www.history.org/foundation/journal/spring04/warfare.cfm.

Graig, Ruth. "Why Did the 1918 Flu Kill So Many Otherwise Healthy Young Adults?" *Smithsonian Magazine*, November 10, 2017. https://www.smithsonianmag.com/history/why-did-1918-flu-kill-so-many-otherwise-healthy-young-adults-180967178/.

Gray, Richard. "Why We Should All Be Wearing Face Masks." *BBC*, May 4, 2020. https://www.bbc.com/future/article/20200504-coronavirus-what-is-the-best-kind-of-face-mask.

"The Great Dying 1616–1619, 'By God's Visitation, a Wonderful Plague.'" Historic Ipswich. Accessed May 1, 2020. https://historicipswich.org/2017/09/01/the-great-dying/.

Harbinger, Jordan. "Are Manmade Viruses the Next Big Terrorist Threat?" *Newsweek*, October 24, 2019. https://www.newsweek.com/2019/11/01/synthetic-biology-manmade-virus-terrorism-1467569.html.

Hoffman, Sarah Z. "HIV/AIDS in Cuba: A Model for Care or an Ethical Dilemma?" *African Health Sciences* 4, no. 3 (December 2004): 208–209. https://www.ncbi.nlm.nih.gov/pmc/articles/PMC2688320/.

"How Do Viruses Evolve?" The Pew Charitable Trusts, March 17, 2020. https://www.pewtrusts.org/en/research-and-analysis/articles/2020/03/17/how-do-viruses-evolve.

"Inside Taiwan's Response to COVID-19." Johns Hopkins. Accessed May 1, 2020. https://www.jhsph.edu/covid-19/events/april-24.html.

Institute of Medicine and National Research Council Committee on Effectiveness of National Biosurveillance Systems. "The BioWatch System." In *Biowatch and Public Health Surveillance: Evaluating Systems for the Early Detection of Biological Threats: Abbreviated Version*. National Academies Press: Washington, DC, 2011. Accessed May 1, 2020. https://www.ncbi.nlm.nih.gov/books/NBK219704.

Jarus, Owen. "20 of the Worst Epidemics and Pandemics in History." Live Science, March 20, 2020. https://www.livescience.com/worst-epidemics-and-pandemics-in-history.html.

Kenen, Joanne. "How Testing Failures Allowed Coronavirus to Sweep the U.S." *Politico*, March 6, 2020. https://www.politico.com/news/2020/03/06/coronavirus-testing-failure-123166.

Koster, John. "Smallpox in the Blankets." HistoryNet, August 2012. https://www.historynet.com/smallpox-in-the-blankets.htm.

LePan, Nicholas. "Visualizing the History of Pandemics." Visual Capitalist, March 14, 2020. https://www.visualcapitalist.com/history-of-pandemics-deadliest/.

Lomas, Natasha. "Israel Passes Emergency Law to Use Mobile Data for COVID-19 Contact Tracing." *Tech Crunch*, March 18, 2020. https://techcrunch.com/2020/03/18/israel-passes-emergency-law-to-use-mobile-data-for-covid-19-contact-tracing/.

Mechanic, Michael. "How to Make a Deadly Pandemic Virus." *Mother Jones*, February 17, 2012. https://www.motherjones.com/environment/2011/12/how-make-deadly-pandemic-virus/.

Molteni, Megan. "Everything You Need to Know About Crispr Gene Editing." *Wired*, December 5, 2017. https://www.wired.com/story/what-is-crispr-gene-editing/.

"A National Blueprint for Biodefense: Leadership and Major Reform Needed to Optimize Efforts." EcoHealth Alliance, October 2015. https://www.ecohealthalliance.org/wp-content/uploads/2016/03/A-National-Blueprint-for-Biodefense-October-2015.pdf.

Nguyen, Hien H. "What Is the Global Incidence of Influenza?" Medscape, January 8, 2020. https://www.medscape.com/answers/219557-3459/what-is-the-global-incidence-of-influenza.

"1918 Pandemic (H1N1 Virus)." CDC, March 20, 2019. https://www.cdc.gov/flu/pandemic-resources/1918-pandemic-h1n1.html.

Nuwer, Rachel. "What If a Deadly Influenza Pandemic Broke Out Today?" *BBC*, November 22, 2018. https://www.bbc.com/future/article/20181120-what-if-a-deadly-influenza-pandemic-broke-out-today.

"Plague Transmission." CDC. Accessed May 1, 2020. https://www.cdc.gov/plague/transmission/index.html.

Powell, Kendall. "How Biologists Are Creating Life-like Cells from Scratch." *Nature*, November 7, 2018. https://www.nature.com/articles/d41586-018-07289-x.

"Prioritizing Diseases for Research and Development in Emergency Contexts." WHO. Accessed May 1, 2020. https://www.who.int/activities/prioritizing-diseases-for-research-and-development-in-emergency-contexts.

"Rainbow Six." Fandom. Accessed May 1, 2020. https://rainbowsix.fandom.com/wiki/Rainbow_Six.

Reid, Rob. "Engineering Superbugs, Accidentally or Otherwise." *Ars Technica*, June 19, 2019. https://arstechnica.com/science/2019/06/ars-on-your-lunch-break-engineering-superbugs-accidentally-or-otherwise/

Reid, Rob. "How Synthetic Biology Could Wipe Out Humanity—and How We Can Stop It." Filmed April 2019 at TED2019, Vancouver, Canada. Video, 16:29.

Reid, Rob. "In the Not-So-distant Future, 'Synbio' Could Lead to Global Catastrophe—Maybe." *Ars Technica*, June 18, 2019. https://arstechnica.com/science/2019/06/what-happens-if-we-engineer-a-superbug-and-the-lab-gets-hacked/.

Roblin, Sebastein. "Fact: Japan Wanted to Drop Plague Bombs on America Using 'Aircraft Carrier' Subs." *The National Interest*, July 17, 2017. https://nationalinterest.org/blog/the-buzz/fact-japan-wanted-drop-plague-bombs-america-using-aircraft-21555.

"Smallpox Factsheet." The Texas Department of Insurance. Accessed May 1, 2020. https://www.tdi.texas.gov/pubs/videoresource/fssmallpox.pdf.

Subramanian, Samanth. "'It's a Razor's Edge We're Walking': Inside the Race to Develop a Coronavirus Vaccine." *The Guardian*, March 27, 2020. https://www.theguardian.com/world/2020/mar/27/inside-the-race-to-develop-a-coronavirus-vaccine-covid-19.

Westcott, Ben, and Shawn Deng. "China Has Made Eating Wild Animals Illegal after the Coronavirus Outbreak. but Ending the Trade Won't Be Easy." *CNN*, March 5, 2020. https://www.cnn.com/2020/03/05/asia/china-coronavirus-wildlife-consumption-ban-intl-hnk/index.html.

"Where Did HIV Come From?" The AIDS Institute. Accessed May 1, 2020. http://www.theaidsinstitute.org/education/aids-101/where-did-hiv-come-0.

Witherspoon, Deborah. "The Most Dangerous Epidemics in U.S. History." Healthline, September 26, 2016. https://www.healthline.com/health/worst-disease-outbreaks-history.

Zimmer, Katarina. "Deforestation Is Leading to More Infectious Diseases in Humans." *National Geographic*, November 22, 2019. https://www.nationalgeographic.com/science/2019/11/deforestation-leading-to-more-infectious-diseases-in-humans/.

CHEMICAL ATTACKS AND ACCIDENTS

"Aum Shinrikyo: The Japanese Cult behind the Tokyo Sarin Attack." *BBC*, July 6, 2018. https://www.bbc.com/news/world-asia-35975069.

Broughton, Edward. "The Bhopal Disaster and Its Aftermath: A Review." *Environmental Health* 4 (May 2005): 6. https://www.ncbi.nlm.nih.gov/pmc/articles/PMC1142333/.

Chabin, Michele. "Israelis Ready Gas Masks amid Syrian Fears." *USA Today*, August 26, 2013. https://www.usatoday.com/story/news/world/2013/08/26/israel-syria-chemical-attacks/2699443/.

Chartrand, Sabra. "Confrontation in the Gulf; Israel's Citizens Receive Gas Masks." *New York Times*, October 8, 1990. https://www.nytimes.com/1990/10/08/world/confrontation-in-the-gulf-israel-s-citizens-receive-gas-masks.html.

Dickinson, Edward, and Love, Jennifer. "A Review of Chemical Warfare Agents and Treatment Options." *Journal of Emergency Medical Services*, September 1, 2017. https://www.jems.com/2017/09/01/a-review-of-chemical-warfare-agents-and-treatment-options/.

Everts, Sarah. "A Brief History of Chemical War." Science History Institute, May 11, 2015. https://www.sciencehistory.org/distillations/a-brief-history-of-chemical-war.

"Facts about Sarin." CDC, April 4, 2018. https://emergency.cdc.gov/agent/sarin/basics/facts.asp.

Franca, Tanos, Daniel A. S. Kitagawa, Samir F. de A. Cavalcante, Jorge A. V. da Silva, Eugenie Nepovimova, and Kamil Kuca. "Novichoks: The Dangerous Fourth Generation of Chemical Weapons." *International Journal of Molecular Sciences* 20, no. 5 (March 2019): 1222. https://www.ncbi.nlm.nih.gov/pmc/articles/PMC6429166/.

"Gas Attack." Montclair State University. Accessed May 1, 2020. https://msu.edu/course/tc/491/internet/students/agnewthesp/sarin/gasattack.html

"Facts About VX." CDC, April 4, 2018. https://emergency.cdc.gov/agent/vx/basics/facts.asp.

"Iraq's Use of Scuds during Operation Desert Storm." Federation of American Scientists. Accessed May 1, 2020. https://fas.org/nuke/guide/iraq/missile/scud_info/scud_info_s04.htm.

Krajick, Kevin. "Defusing Africa's Killer Lakes." *Smithsonian Magazine*, September 2003. https://www.smithsonianmag.com/science-nature/defusing-africas-killer-lakes-88765263/.

"New Technologies, Artificial Intelligence Aid Fight against Global Terrorism." *UN News*, September 4, 2019. https://news.un.org/en/story/2019/09/1045562.

"1991: Iraqi Scud Missiles Hit Israel." *BBC*. Accessed May 1, 2020. http://news.bbc.co.uk/onthisday/hi/dates/stories/january/18/newsid_4588000/4588486.stm.

Pockett, Consuella. "United States and Israeli Homeland Security: A Comparative Analysis of Emergency Preparedness Efforts." US Air Force Counterproliferation Center, August 2005. https://media.defense.gov/2019/Apr/11/2002115499/-1/-1/0/33USANDISREALICOMPARE.PDF.

"Preparing for a Terror Attack." *Chicago Tribune*, February 13, 2003. https://www.chicagotribune.com/news/ct-xpm-2003-02-13-0302130212-story.html.

"Scuds and Chemical and Biological Weapons." Federation of American Scientists. Accessed May 1, 2020. https://fas.org/nuke/guide/iraq/missile/scud_info/scud_info_s03.htm.

Sever, M., Y. Garb, and D. Pearlmutter. "Building in Resilience: Long-Term Considerations in the Design and Production of Residential Buildings in Israel." *Disaster Management: Enabling Resilience*, edited by Anthony Masys, 65–91. Springer: New York, 2014.

Sorensen, John, and Barbara Vogt. "Will Duct Tape and Plastic Really Work? Issues Related to Expedient Shelter-in-Place." Federation of American Scientists, August 2001. https://fas.org/irp/threat/duct.pdf.

Taylor, Alan. "Bhopal: The World's Worst Industrial Disaster, 30 Years Later." *The Atlantic*, December 2, 2014. https://www.theatlantic.com/photo/2014/12/bhopal-the-worlds-worst-industrial-disaster-30-years-later/100864/.

"What Is the UK-Invented VX Nerve Agent—and Why Is It So Deadly?" *The Telegraph*, August 13, 2017. https://www.telegraph.co.uk/news/2017/02/24/vx-nerve-agent-deadly/.

Zion, Ilan Ben. "Vital Sarin Antidote Missing from Gas Mask Kits." *The Times of Israel*, August 29, 2013. https://www.timesofisrael.com/gas-mask-kits-contain-no-antidote-for-nerve-gas/.

GRID ATTACK

Amatulli, Jenna, and Hayley Miller. "Puerto Ricans Describe 'Horror in the Streets' after Hurricane Maria." *Huffington Post*, September 25, 2017. https://www.huffpost.com/entry/puerto-rico-hurricane-maria_n_59c90dcfe4b0cdc77332e283.

Bizjak, Tony. "Day 2 of PG&E Blackout: Over 1.5 Million Californians in the Dark, Power Back for Some." The Sacramento Bee, October 11, 2019. https://www.sacbee.com/news/california/article235976801.html.

Brumfield, Cynthia. "Regional Municipal Ransomware Attacks Soar; MS-ISAC Can Help." CSO, August 28, 2019. https://www.csoonline.com/article/3433930/regional-municipal-ransomware-attacks-soar-ms-isac-can-help.html.

Chabria, Anita, and Luna, Taryn. "PG&E Power Outages Bring Darkness, Stress and Debt to California's Poor and Elderly." *Los Angeles Times*, October 11, 2019. https://www.latimes.com/california/story/2019-10-11/pge-power-outage-darkness-stress-debt-vulnerable.

Fuks, Lena. "10 Ransomware Attacks You Should Know About in 2019." Allot, April 28, 2019. https://www.allot.com/blog/10-ransomware-attacks-2019/.

Harte, Julia. "Trump Tweet, Political Divisions Fuel Rising Discourse about New U.S. Civil War." *Reuters*, October 29, 2019. https://www.reuters.com/article/us-usa-civil-war/trump-tweet-political-divisions-fuel-rising-discourse-about-new-us-civil-war-idUSKBN1X812B.

"How the Electricity Grid Works." Union of Concerned Scientists, February 17, 2015. https://www.ucsusa.org/clean-energy/how-electricity-grid-works.

Maxouris, Christina. "On the Exact Same Day 42 Years Ago, a New York Power Outage Turned into a Crime Rampage." *CNN*, July 14, 2019. https://www.cnn.com/2019/07/14/us/new-york-city-power-outage-42-years-trnd/index.html.

Minkel, J. R. "The 2003 Northeast Blackout—Five Years Later." *Scientific American*, August 13, 2008. https://www.scientificamerican.com/article/2003-blackout-five-years-later/.

Moench, Mallory. "PG&E Outages: Long Lines for Gas, Stations Are Closing." *San Francisco Chronicle*, October 9, 2019. https://www.sfchronicle.com/business/article/Gas-another-worry-as-PG-E-shuts-off-power-14504278.php.

Occupational Safety and Health Administration. "Illustrated Glossary: Substations." United States Department of Labor. Accessed May 1, 2020. https://www.osha.gov/SLTC/etools/electric_power/illustrated_glossary/substation.html.

Phillips, Morgan. "Oxygen-Dependent California Man Dies 12 Minutes after PG&E Cuts Power to His Home." *Fox News*, October 11, 2019. https://www.foxnews.com/us/oxygen-dependent-man-dies-12-minutes-after-pge-cuts-power-to-his-home.

"Saboteurs behind GRIDCo Tower Hack—Energy Minister." *GH Headlines*, March 25, 2019. http://www.ghheadlines.com/agency/3news/20190325/114962461/saboteurs-behind-gridco-tower-hack-energy-minister.

Shinhoe, Dan. "The 15 Biggest Data Breaches of the 21st Century." CSO Online, April 17, 2020. https://www.csoonline.com/article/2130877/the-biggest-data-breaches-of-the-21st-century.html.

Sobczak, Blake. "Report Reveals Play-by-Play of First U.S. Grid Cyberattack." E&E News, September 6, 2019. https://www.eenews.net/stories/1061111289.

"U.S. Electric System Is Made Up of Interconnections and Balancing Authorities." U.S. Energy Information Administration, July 20, 2016. https://www.eia.gov/todayinenergy/detail.php?id=27152.

"What Are the Requirements for Hospital Backup Generators?" Woodstock Power Company. Accessed May 1, 2020. https://woodstockpower.com/blog/what-are-the-requirements-for-hospital-backup-generators/.

THE AUTOMATION ECONOMY

Adiwardana, Daniel, Minh-Thang Luong, David So, Jamie Hall, Noah Fiedel, Romal Thoppilan, Zi Yang, Apoory Kulshreshtha, Gauray Nemade,Yifeng Lu , and Quoc Le. "Towards a Human-like Open-Domain Chatbot." Cornell University, January 27, 2020. https://arxiv.org/abs/2001.09977.

Al Jazeera English. "The Dark Side of Bangladesh leather." YouTube video, 2:21. December 2, 2013. https://www.youtube.com/watch?v=crfnmrKexzs.

Bain, Marc. "A New T-Shirt Sewing Robot Can Make as Many Shirts per Hour as 17 Factory Workers." *Quartz*, August 30, 2017. https://qz.com/1064679/a-new-t-shirt-sewing-robot-can-make-as-many-shirts-per-hour-as-17-factory-workers/.

Bain, Mark, and Jenni Avins. "The Thing That Makes Bangladesh's Garment Industry Such a Huge Success Also Makes It Deadly." *Quartz*, April 24, 2015. https://qz.com/389741/the-thing-That-makes-bangladeshs-garment-industry-such-a-huge-success-also-makes-it-deadly/.

Bedord, Laurie. "How Automation Will Transform Farming." *Successful Farming*, November 29, 2017. https://www.agriculture.com/technology/robotics/how-automation-will-transform-farming.

Bellis, Mary. "History of American Agriculture." Thought Co., July 3, 2019. https://www.thoughtco.com/history-of-american-agriculture-farm-machinery-4074385.

Best, Jo. "IBM Watson: The Inside Story of How the Jeopardy-Winning Supercomputer Was Born, and What it Wants to Do Next." *Tech Republic*, September 9, 2013. https://www.techrepublic.com/article/ibm-watson-the-inside-story-of-how-the-jeopardy-winning-supercomputer-was-born-and-what-it-wants-to-do-next/.

Blumberg, Julie, and Gabriel, Kreiman. "How Cortical Neurons Help Us See: Visual Recognition in the Human Brain." *Journal of Clinical Investigation* 120, no. 9 (September 2010): 3054–3063. https://www.ncbi.nlm.nih.gov/pmc/articles/PMC2929717/.

"Brain Facts and Figures." University of Washington. Accessed May 1, 2020. https://faculty.washington.edu/chudler/facts.html.

Bryant, Ben. "Judges Are More Lenient after Taking a Break, Study Finds." *The Guardian*, April 11, 2011. https://www.theguardian.com/law/2011/apr/11/judges-lenient-break.

Butler, Sarah. "Why Are Wages So Low for Garment Workers in Bangladesh?" *The Guardian*, January 21, 2019. https://www.theguardian.com/business/2019/jan/21/low-wages-garment-workers-bangladesh-analysis.

Carter, Terry. "Implicit Bias Is a Challenge Even for Judges." *ABA Journal*, August 5, 2016. https://www.abajournal.com/news/article/implicit_bias_is_a_challenge_even_for_judges.

Carver, Nicki, Vikas Gupta, and John Hipskind. "Medical Error." National Center for Biotechnology Information, February 16, 2020. https://www.ncbi.nlm.nih.gov/books/NBK430763/.

The Daily Star. "A Day in the Life of a Minimum Wage Earner in Bangladesh." YouTube video, 3:35. April 3, 2017. https://www.youtube.com/watch?v=8QSC_9c6qCQ.

Daniel, Micheal. "Study Suggests Medical Errors Now Third Leading Cause of Death in the U.S." Johns Hopkins Medicine, May 3, 2016. https://www.hopkinsmedicine.org/news/media/releases/study_suggests_medical_errors_now_third_leading_cause_of_death_in_the_us.

Deutscher, Maria. "Challenging Nvidia, Intel Unleashes Three New Chips for AI Work." SiliconAngle, November 12, 2019. https://siliconangle.com/2019/11/12/challenging-nvidia-intel-unleashes-new-chips-ai-training-inference/.

Elliot, Larry. "World's 26 Richest People Own as Much as Poorest 50%, Says Oxfam." *The Guardian*, January 20, 2019. https://www.theguardian.com/business/2019/jan/21/world-26-richest-people-own-as-much-as-poorest-50-per-cent-oxfam-report.

Feroldi. Brian. "7 Near-Monopolies That Are Perfectly Legal in America." Motley Fool, July 21, 2017. https://www.fool.com/investing/2017/07/21/7-near-monopolies-That-are-perfectly-legal-in-amer.aspx.

Fisher, Eve. "The $3500 Shirt—A History Lesson in Economics." Sleuth Sayers, June 6, 2013. https://www.sleuthsayers.org/2013/06/the-3500-shirt-history-lesson-in.html.

Fisher, Max. "Here's a Map of the Countries That Provide Universal Health Care (America's Still Not on It)." *The Atlantic*, June 28, 2012. https://www.theatlantic.com/international/archive/2012/06/heres-a-map-of-the-countries-That-provide-universal-health-care-americas-still-not-on-it/259153/.

Frank, Robert. "Billionaires Hurt Economic Growth and Should Be Taxed Out of Existence, Says Bestselling French Economist." *CNBC*, September 12, 2019. https://www.cnbc.com/2019/09/12/billionaires-should-be-taxed-out-of-existence-says-thomas-piketty.html.

Gold, Howard. "Never Mind the 1 Percent: Let's Talk about the 0.01 Percent." Chicago Booth Review. Accessed May 1, 2020. https://review.chicagobooth.edu/economics/2017/article/never-mind-1-percent-lets-talk-about-001-percent.

Guilbert, Kieran. "Adidas, Nike Urged to Ensure Fair Wages for Asian Workers Making World Cup Kits." Reuters, June 11, 2018. https://www.reuters.com/article/us-asia-workers-worldcup/adidas-nike-urged-to-ensure-fair-wages-for-asian-workers-making-world-cup-kits-idUSKBN1J727J.

Ingram, Christiphor. "U.N. Warns That Runaway Inequality Is Destabilizing the World's Democracies." Washington Post, February 11, 2020. https://www.washingtonpost.com/business/2020/02/11/income-inequality-un-destabilizing/.

Jacobs, Harrison. "Inside 'iPhone City,' the Massive Chinese Factory Town Where Half of the World's iPhones Are Produced." Business Insider, May 7, 2018. https://www.businessinsider.com/apple-iphone-factory-foxconn-china-photos-tour-2018-5.

Kotkin, Joel. "How America Is Reverting Back to the Feudal Age." New York Post, December 25, 2019. https://nypost.com/2019/12/25/how-america-is-reverting-back-to-the-feudal-age/.

Kochhar, Rakesh. "Seven-in-Ten People Globally Live on $10 or Less per Day." Pew Research Center, September 23, 2015. https://www.pewresearch.org/fact-tank/2015/09/23/seven-in-ten-people-globally-live-on-10-or-less-per-day/.

Kuper, Simon. "This Economist Has a Radical Plan to Solve Wealth Inequality." Wired, April 14, 2020. https://www.wired.co.uk/article/thomas-piketty-capital-ideology.

Luscombe, Richard. "Life Expectancy Gap between Rich and Poor US Regions Is 'More Than 20 Years.'" The Guardian, May 8, 2017. https://www.theguardian.com/inequality/2017/may/08/life-expectancy-gap-rich-poor-us-regions-more-than-20-years.

Martin, Lisa. "Workers Making Clothes for Australian Brands Can't Afford to Eat, Oxfam Reports." The Guardian, February 24, 2019. https://www.theguardian.com/fashion/2019/feb/25/australian-fashion-brand-workers-earning-51-cents-an-hour-oxfam-reports.

Mishel, Lawrence, and Jessica Schiedera. "CEO Compensation Surged in 2017." Economic Policy Institute, August 16, 2018. https://www.epi.org/publication/ceo-compensation-surged-in-2017/.

Moore, Sam. "Miles per Acre." Farm Collector, March 4, 2009. https://www.farmcollector.com/farm-life/miles-per-acre.

Orca, Surfdaddy. "Brain on a Chip." Humanity Plus Magazine, April 7, 2009. https://hplusmagazine.com/2009/04/07/brain-chip/.

Pichhi, Aimee. "It Now Takes Up to 66 weeks to Pay for 52 weeks of Middle-class Basics." CBS News, February 27, 2020. https://www.cbsnews.com/news/a-thriving-middle-class-life-requires-more-than-a-years-income/.

Reddy, K. P. "SoftWear Automation Launches LOWRY Advanced Sewing Robot Line." Advanced Technology Development Center, October 26, 2015. https://atdc.org/news-from-our-companies/softwear-automation-launches-lowry-advanced-sewing-robot-line/.

Seabrook, John. "The Age of Robot Farmers." New Yorker, April 8, 2019. https://www.newyorker.com/magazine/2019/04/15/the-age-of-robot-farmers.

Teather, David. "Nike Lists Abuses at Asian Factories." The Guardian, April 14, 2005. https://www.theguardian.com/business/2005/apr/14/ethicalbusiness.money.

Thomson Reuters Foundation. "Bangladeshi Garment Worker Seeks Better Life for Daughters." YouTube video, 3:45. August 13, 2013. https://www.youtube.com/watch?v=17Uctg_2GPE.

Urdaneta, Sheyla. "Venezuelan City Devastated by Looting during Power Outages." Associated Press, March 13, 2019. https://apnews.com/05173298ce6d42f58691791c2a134ca4.

Vellacott, Chris. "Super Rich Hold $32 Trillion in Offshore Havens." Reuters, July 22, 2012. https://www.reuters.com/article/usoffshore-wealth-idUSBRE86L03U20120722.

Weber-Steinhaus, Fiona. "The Rise and Rise of Bangladesh—but Is Life Getting Any Better?" The Guardian, October 9, 2019. https://www.theguardian.com/global-development/2019/oct/09/bangladesh-women-clothes-garment-workers-rana-plaza.

"Why 80% of Singaporeans Live in Government-Built Flats." The Economist, July 6, 2017. https://www.economist.com/asia/2017/07/06/why-80-of-singaporeans-live-in-government-built-flats.

Wishart, David. "Oakies." Plains Humanities. Accessed May 1, 2020. http://plainshumanities.unl.edu/encyclopedia/doc/egp.ii.044.xml.

OPIOID CRISIS

Al-Hasani, Ream, and Bruchas, Michael. "Molecular Mechanisms of Opioid Receptor-Dependent Signaling and Behavior." Anesthesiology 115, no. 6 (December 2011): 1363–1381. https://www.ncbi.nlm.nih.gov/pmc/articles/PMC3698859/.

"America's Drug Overdose Epidemic: Data to Action." CDC, March 24, 2020. https://www.cdc.gov/injury/features/prescription-drug-overdose/index.html.

"Bayer: A History." GMWatch. Accessed, May 1, 2020. https://www.gmwatch.org/en/articles/gm-firms/bayer-a-history.

Berridge, Kent, and Morten Kringelbach. "Pleasure Systems in the Brain." Neuron 86, no. 3 (May 2015): 646–664. https://www.ncbi.nlm.nih.gov/pmc/articles/PMC4425246/.

Chiasson, Dan. "The Man Who Invented the Drug Memoir." New Yorker, October 10, 2016. https://www.newyorker.com/magazine/2016/10/17/the-man-who-invented-the-drug-memoir.

De Quincy, Thomas. "Confessions of an English Opium-Eater." London Magazine, September 1821. https://www.gutenberg.org/files/2040/2040-h/2040-h.htm.

DeWeerdt, Sarah. "Tracing the US Opioid Crisis to Its Roots." Nature, September 11, 2019. https://www.nature.com/articles/d41586-019-02686-2.

Durbin, Kaci. "Oxycodone." Drugs.com, November 4, 2019. https://www.drugs.com/oxycodone.html.

"Fentanyl." PubChem. Accessed May 1, 2020. https://pubchem.ncbi.nlm.nih.gov/compound/Fentanyl.

Ferreira, Susana. "Portugal's Radical Drugs Policy Is Working. Why Hasn't the World Copied It?" The Guardian, December 5, 2017. https://www.theguardian.com/news/2017/dec/05/portugals-radical-drugs-policy-is-working-why-hasnt-the-world-copied-it.

Fox, Maggie. "Death Maps Show Where Despair Is Killing Americans." NBC News, March 13, 2018. https://www.nbcnews.com/health/health-news/death-maps-show-where-despair-killing-americans-n856231.

Hartmann, Thom. "It's Time to Bring Back the Corporate Death Penalty." Common Dreams, January 8, 2019. https://www.commondreams.org/views/2019/01/08/its-time-bring-back-corporate-death-penalty.

Hayhurst, Christina and Durieux, Marcel. "Opioid Overdose: The Price of Tolerance." Scientific American, August 9, 2016. https://blogs.scientificamerican.com/guest-blog/opioid-overdose-the-price-of-tolerance/.

Higham, Scott, Sari Horwitz, and Steven Rich. "76 Billion Opioid Pills: Newly Released Federal Data Unmasks the Epidemic." Washington Post, July 16, 2019. https://www.washingtonpost.com/investigations/76-billion-opioid-pills-newly-released-federal-data-unmasks-the-epidemic/2019/07/16/5f29fd62-a73e-11e9-86dd-d7f0e60391e9_story.html.

"How Addiction Hijacks the Brain." Harvard Health Publishing, July, 2011. https://www.health.harvard.edu/newsletter_article/how-addiction-hijacks-the-brain.

"How Opioid Drugs Activate Receptors." National Institutes of Health, May 22, 2018. https://www.nih.gov/news-events/nih-research-matters/how-opioid-drugs-activate-receptors.

Khazan, Olga. "The Opioid Epidemic Might Be Much Worse Than We Thought." *The Atlantic,* February 27, 2020. https://www.theatlantic.com/health/archive/2020/02/more-people-have-died-opioids-us-thought/607165/.

Kline, Thomas, and Max Lamb. "SUICIDES Associated with Forced Tapering of Opiate Pain Treatments." Medium, May 11, 2018. https://medium.com/@ThomasKlineMD/opioidcrisis-pain-related-suicides-associated-with-forced-tapers-c68c79ecf84d.

Lopez, German. "White House: The Opioid Epidemic Cost $2.5 Trillion Over 4 Years." *Vox*, November 1, 2019. https://www.vox.com/policy-and-politics/2019/11/1/20943599/opioid-epidemic-cost-white-house-economic-advisers.

McGreal, Chris. "Why Were Millions of Opioid Pills Sent to a West Virginia Town of 3,000?" *The Guardian*, October 2, 2019. https://www.theguardian.com/us-news/2019/oct/02/opioids-west-virginia-pill-mills-pharmacies.

"Opioid Overdose Crisis." National Institute on Drug Abuse, April, 2020. https://www.drugabuse.gov/drugs-abuse/opioids/opioid-overdose-crisis/.

"Opium Poppy." DEA Museum. Accessed May 1, 2020. https://www.deamuseum.org/ccp/opium/history.html.

"Overdose Death Rates." National Institute on Drug Abuse. Accessed May 1, 2020. https://www.drugabuse.gov/related-topics/trends-statistics/overdose-death-rates.

"Overprescribing Opioids." Washington University School of Medicine in St. Louis, April 27, 2016. https://medicine.wustl.edu/news/podcast/overprescribing-opioids/.

Pappas, Stephanie. "Massive Poppy Bust: Why Home-Grown Opium Is Rare." Live Science, June 12, 2017. https://www.livescience.com/59452-why-opium-is-grown-outside-us.html.

Ronald, Hirsch. "The Opioid Epidemic: It's Time to Place Blame Where It Belongs." *Missouri Medicine* 114, no. 2 (March–April 2017): 82–83, 90. https://www.ncbi.nlm.nih.gov/pmc/articles/PMC6140023/.

Ryan, Harriet, Lisa Girion, and Scott Glover. "'You Want a Description of Hell?' OxyContin's 12-Hour Problem." *Los Angeles Times*, May 5, 2016. https://www.latimes.com/projects/oxycontin-part1/.

Shah, Mansi, and Martin Huecker. "Opioid Withdrawal." National Center for *StatPearls*. StatPearls Publishing: Treasure Island, Florida, 2019. https://www.ncbi.nlm.nih.gov/books/NBK526012/.

Sidarth, Wakhlu. "From Pain Pills to Heroin: How We Can Save Patients from the Slippery Slope." UT Southwestern Medical Center, May 30, 2018. https://utswmed.org/medblog/opioid-addiction-treatment/.

"Study: 61% of Opioid-Related Deaths Linked to Chronic Pain Diagnosis." November 29, 2017, American Physical Therapy Association. https://www.apta.org/PTinMotion/News/2017/11/29/OpioidDeathsAndChronicPain/.

Van Zee, Art. "The Promotion and Marketing of OxyContin: Commercial Triumph, Public Health Tragedy." *American Journal of Public Health* 99, no. 2 (February 2009): 221–227. https://www.ncbi.nlm.nih.gov/pmc/articles/PMC2622774/.

Wang, Dong, Gregory Scherrer, and Elizabeth Sypek. "Researchers Identify Source of Opioids' Side Effects." Stanford Medicine, January 16, 2017. https://med.stanford.edu/news/all-news/2017/01/commercial-drug-found-to-limit-opioids-side-effects-in-mice.html.

"What Effects Does Heroin Have on the Body?" National Institute on Drug Abuse, June, 2018. https://www.drugabuse.gov/publications/research-reports/heroin/how-heroin-used.

ASTEROID STRIKE

"Asteroid Danger Explained." European Space Agency, June 21, 2018. https://www.esa.int/ESA_Multimedia/Images/2018/06/Asteroid_danger_explained.

Blašković, Teo. "Large Asteroid 2018 AH Flew Past Earth at 0.77 LD, 2 Days before Discovery." The Watchers, January 8, 2018. https://watchers.news/2018/01/08/asteroid-2018-ah/.

"Blast Effects on Humans." Atomic Archive. May 1, 2020. http://www.atomicarchive.com/Effects/effects5.shtml.

"Chemical Explosives." Federation of American Scientists. Accessed May 1, 2020. https://fas.org/man/dod-101/navy/docs/es310/chemstry/chemstry.htm.

Chiu, Allyson. "'It Snuck Up on Us': Scientists Stunned by 'City-Killer' Asteroid That Just Missed Earth." *Washington Post*, July 26, 2019. https://www.washingtonpost.com/nation/2019/07/26/it-snuck-up-us-city-killer-asteroid-just-missed-earth-scientists-almost-didnt-detect-it-time/.

Denton, Adeene. "The Venus Controversy." The Planetary Society, August 14, 2018. http://www.planetary.org/blogs/guest-blogs/the-venus-controversy.html.

"Discovering the Impact Site." Lunar and Planetary Institute. Accessed May 1, 2020. https://www.lpi.usra.edu/science/kring/Chicxulub/discovery/.

Donahue, Michelle. "Dino-Killing Asteroid Hit Just the Right Spot to Trigger Extinction." *National Geographic*, November 9, 2017. https://www.nationalgeographic.com/news/2017/11/dinosaurs-extinction-asteroid-chicxulub-soot-earth-science/.

Kornei, Katherine. "Huge Global Tsunami Followed Dinosaur-Killing Asteroid Impact." Eos, December 20, 2018. https://eos.org/articles/huge-global-tsunami-followed-dinosaur-killing-asteroid-impact.

Letzter, Rafi. "An Asteroid with Its Own Moon Will Zip Past Earth Tonight." Space.com, May 25, 2019. https://www.space.com/asteroid-passes-close-to-earth.html.

McKinnon, Mika, and Misra, Ria. "A Scientists Responds . . . to Deep Impact." *Gizmodo*, June 10, 2015. https://io9.gizmodo.com/a-scientist-responds-to-deep-impact-1709206458.

"Nuclear Detonation: Weapons, Improvised Nuclear Devices." Radiation Emergency Medical Management. Accessed May 1, 2020. https://www.remm.nlm.gov/nuclearexplosion.htm.

O'Neill, Ian. "If an Asteroid Hits the Ocean, Does It Make a Tsunami? (Probably Not)." Space.com, December 19, 2016. https://www.space.com/35081-asteroid-impact-ocean-computer-simulations-solar-system.html.

Oskin, Becky. "Crash! 10 Biggest Impact Craters on Earth." Live Science, April 28, 2014. https://www.livescience.com/45126-biggest-impact-crater-earth-countdown.html.

"Planetary Defense Frequently Asked Questions." NASA. Accessed May 1, 2020. https://www.nasa.gov/planetarydefense/faq.

Seeker. "An Asteroid Didn't Kill the Dinosaurs, Here's a New Theory About What Did." YouTube video, 2:14. September 23, 2017. https://www.youtube.com/watch?v=VrbqkeB4UqE.

SUPERVOLCANO ERUPTION

Bagley, Mary. "Mount St. Helens Eruptions: Facts & Information." Live Science, October 16, 2018. https://www.livescience.com/27553-mount-st-helens-eruption.html.

Gitlin, Jonathan M. "Krakatoa's Chilling Effect." *Ars Technica*, February 9, 2006. https://arstechnica.com/science/2006/02/2815/.

Hall, Shannon. "Yellowstone Supervolcano Could Be an Energy Source. but Should It?" *National Geographic*, August 8, 2018. https://www.nationalgeographic.com/science/2018/08/news-yellowstone-supervolcano-geothermal-energy-debate-iceland-hawaii/.

"How Do the Giant Eruptions in the Yellowstone National Park Region Compare to Other Large Historic Eruptions?" United States Geological Survey. Accessed May 1, 2020. https://www.usgs.gov/faqs/how-do-giant-eruptions-yellowstone-national-park-region-compare-other-large-historic-eruptions.

Martin, Sean. "Supervolcanoes MAPPED: Where the World's Biggest Volcanoes Are Which Could End ALL LIFE." Express, August 17, 2018. https://www.express.co.uk/news/science/1004358/supervolcano-map-yellowstone-long-valley-caldera-lake-toba-taupo-campi-flegri.

"Mount Tambora and the Year Without a Summer." UCAR Center for Science Education. Accessed May 1, 2020. https://scied.ucar.edu/shortcontent/mount-tambora-and-year-without-summer.

National Geographic. "Supervolcanoes 101." YouTube video, 3:40. August 6, 2018. https://www.youtube.com/watch?v=kAlawvE8lVw.

Nature video. "Pyroclastic flows: The Secret of Their Deadly Speed." YouTube video, 2:15. April 8, 2019. https://www.youtube.com/watch?v=hvuP7kuX7Dk.

Reddy, Prishani. "Differences between the Earths' Lithosphere and Asthenosphere." Differencebetween.net. Accessed May 1, 2020. http://www.differencebetween.net/science/differences-between-the-earths-lithosphere-and-asthenosphere/.

"Ring of Fire." National Geographic. Accessed May 1, 2020. https://www.nationalgeographic.org/encyclopedia/ring-fire/.

Seeker. "NASA's Crazy Plan to Save the World from the Supervolcano Under Yellowstone." YouTube video, 3:32. November 7, 2007. https://www.youtube.com/watch?v=wt_r5xO2JkA.

Smithsonian Channel. "Why the Yellowstone Supervolcano Could Be Huge." YouTube video, 3:29. June 5, 2015. https://www.youtube.com/watch?v=lMLo0E66O8A.

TED-ed. "The colossal consequences of supervolcanoes—Alex Gendler." YouTube video, 4:50. June 9, 2014. https://www.youtube.com/watch?v=hDNlu7Qf6_E.

"What Is a Hot Spot?" Oregon State University. Accessed May 1, 2020. http://volcano.oregonstate.edu/what-is-a-hot-spot.

EARTHQUAKES

Bilek, Susan L., and Throne Lay. "Subduction Zone Megathrust Earthquakes." Geo Science World, July 6, 2018. https://pubs.geoscienceworld.org/gsa/geosphere/article/14/4/1468/541663/Subduction-zone-megathrust-earthquakes.

Brinklow, Adam. "The Bay Area's Biggest Earthquakes." Curbed San Francisco, October 18, 2019. https://sf.curbed.com/maps/map-bay-area-biggest-earthquakes-1906-loma-prieta-san-francisco.

Dove, Adam. "Why Has Oklahoma Experienced a Nearly 4000% Increase in Earthquake Activity?" Phys.org, November 21, 2016. https://phys.org/news/2016-11-oklahoma-experienced-earthquake.html.

The Editors of Encyclopedia Britannica. "Ring of Fire." Encyclopedia Britannica. Accessed May 1, 2020. https://www.britannica.com/place/Ring-of-Fire.

Holzer, Thomas. "The 1995 Hanshin-Awaji (Kobe), Japan, Earthquake" GSA Today 5, no. 8 (August 1995): 153–156, 165. https://www.geosociety.org/gsatoday/archive/5/8/pdf/i1052-5173-5-8-sci.pdf.

"Latest Earthquakes." United States Geological Survey. https://earthquake.usgs.gov/earthquakes/map/.

Lynch, David K. "The San Andreas Fault." Geology.com. Accessed May 1, 2020. https://geology.com/articles/san-andreas-fault.shtml.

National Geographic. "Earthquakes 101." YouTube video, 5:01. January 23, 2020. https://www.youtube.com/watch?v=_r_nFT2m-Vg.

New Media Communications at OSU. "Cascadia Subduction Zone: The Big One." YouTube video, 7:58. June 29, 2016. https://www.youtube.com/watch?v=e6U198ULMYo.

"1906 Earthquake Fence." Atlas Obscura. Accessed May 1, 2020. https://www.atlasobscura.com/places/earthquake-fence-skb.

Oskin, Becky. "What Is Plate Techtonics?" Live Science, December 19, 2017. https://www.livescience.com/37706-what-is-plate-tectonics.html.

Phillips, Campbell. "Earthquakes: The 10 Biggest in History." Australian Geographic, March 14, 2011. https://www.australiangeographic.com.au/topics/science-environment/2011/03/earthquakes-the-10-biggest-in-history/.

Simpson Strong-Tie. "World's Largest Earthquake Test." YouTube video, 2:27. June 21, 2011. https://www.youtube.com/watch?v=hSwjkG3nv1c.

"Understanding Plate Motions." United States Geological Survey, September 15, 2014. https://pubs.usgs.gov/gip/dynamic/understanding.html.

vos Savant, Marilyn. "Is a 10.0 Magnitude Earthquake Possible?" Parade, June 12, 2016. https://parade.com/482222/marilynvossavant/is-a-10-0-magnitude-earthquake-possible/.

Wald, Lisa. "The Science of Earthquakes." United States Geological Survey. Accessed May 1, 2020. https://www.usgs.gov/natural-hazards/earthquake-hazards/science/science-earthquakes?qt-science_center_objects=0#qt-science_center_objects.

SUPERTSUNAMIS

Geist, Eric. "Life of a Tsunami." United States Geological Survey. Accessed May 1, 2020. https://www.usgs.gov/centers/pcmsc/science/life-a-tsunami?qt-science_center_objects=0#qt-science_center_objects.

Hambling, David. "The Truth Behind Russia's 'Apocalypse Torpedo.'" Popular Mechanics, January 18, 2019. https://www.popularmechanics.com/military/weapons/a25953089/russia-apocalypse-torpedo-poseidon/.

Jones, Anna. "Indonesia Tsunami: Palu Hit by 'Worst Case Scenario.'" BBC, October 2, 2018. https://www.bbc.com/news/world-asia-45702566.

Kyung-hoon, Kim. "After the Tsunami: Japan's Sea Walls—in Pictures." The Guardian, March 9, 2018. https://www.theguardian.com/world/gallery/2018/mar/09/after-the-tsunami-japan-sea-walls-in-pictures.

Lockie, Alex. "The Real Purpose of Russia's 100-Megaton Underwater Nuclear Doomsday Device." Business Insider, February 11, 2019. https://www.businessinsider.com/the-real-purpose-of-russias-poseidon-nuclear-doomsday-device-2019-2.

Lockie, Alex. "Why Putin's New 'Doomsday' Device Is So Much More Deadly and Horrific Than a Regular Nuke." Business Insider, March 15, 2018. https://www.businessinsider.com/putin-doomsday-status-6-nuclear-weapon-2018-3.

Mearian, Lucas. "Memory Chip Prices Surge in Aftermath of Japan's Quake." Computer World, March 14, 2011. https://www.computerworld.com/article/2506919/memory-chip-prices-surge-in-aftermath-of-japan-s-quake.html.

Mosher, Dave. "A New Russian Video May Show a 'Doomsday Machine' Able to Trigger 300-Foot Tsunamis—but Nuclear Weapons Experts Question Why You'd Ever Build One." Business Insider, July 24, 2018. https://www.businessinsider.com/russia-doomsday-weapon-submarine-nuke-2018-4.

National Geographic. "Rare Video: Japan Tsunami." YouTube video, 3:34. June 13, 2011. https://www.youtube.com/watch?v=oWzdgBNfhQU.

Oskin, Becky. "Japan Earthquake & Tsunami of 2011: Facts and Information." Live Science, September 13, 2017. https://www.livescience.com/39110-japan-2011-earthquake-tsunami-facts.html.

Phillips, Campbell. "The 10 Most Destructive Tsunamis in History." Australian Geographic, March 16, 2011. https://www.australiangeographic.com.au/topics/science-environment/2011/03/the-10-most-destructive-tsunamis-in-history/.

"Timeline: 100 Years of Deadly Tsunamis." ABC News, October 26, 2010. https://www.abc.net.au/news/2004-12-29/timeline-100-years-of-deadly-tsunamis/610126.

"Tsunamis: Facts About Killer Waves." National Geographic, January 14, 2005. https://www.nationalgeographic.com/news/2005/1/tsunamis-facts-about-killer-waves/.

"What Happened during the 2004 Sumatra Earthquake." Tectonics Observatory. Accessed May 1, 2020. http://www.tectonics.caltech.edu/outreach/highlights/sumatra/what.html.

Wren, G. G., and D. May. "Detection of Submerged Vessels Using Remote Sensing Techniques." *Australian Defense Force Journal* no. 127 (December 1997): 9–15. https://fas.org/nuke/guide/usa/slbm/detection.pdf.

CORONAL MASS EJECTIONS

"The Aurora Borealis." *New York Times*, May 29, 1877. http://www.solarstorms.org/NewsPapers/1877a.pdf.

Bennett, Joe. "What Is a Geomagnetic Disturbance and How Can It Affect the Power Grid?" North American Energy Services, December 10, 2017. https://www.naes.com/news/what-is-a-geomagnetic-disturbance-and-how-can-it-affect-the-power-grid/.

Boyle, Rebecca. "How We'll Safeguard Earth from a Solar Storm Catastrophe." *NBC News*, June 8, 2017. https://www.nbcnews.com/mach/space/how-we-ll-safeguard-earth-solar-storm-catastrophe-n760021.

Brundige, Ellen. "If the Massive Solar Flare of 1859 (The "Carrington Flare") Happened Today . . ." Owlcation, January 11, 2018. https://owlcation.com/stem/massive-solar-flare-1859.

Garner, Rob. "Understanding the Magnetic Sun." NASA, January 29, 2016. https://www.nasa.gov/feature/goddard/2016/understanding-the-magnetic-sun.

Howell, Elizabeth. "Giant Halloween Solar Storm Sparked Earth Scares 10 Years Ago (Video)." Space.com, October 30, 2013. https://www.space.com/23396-scary-halloween-solar-storm-2003-anniversary.html.

Lovett, Richard A. "What If the Biggest Solar Storm on Record Happened Today?" *National Geographic*, March 4, 2011. https://www.nationalgeographic.com/news/2011/3/110302-solar-flares-sun-storms-earth-danger-carrington-event-science/.

Mann, Adam. "1 in 8 Chance of Catastrophic Solar Megastorm by 2020." *Wired*, February 29, 2012. https://www.wired.com/2012/02/massive-solar-flare/.

North American Electric Reliability Corporation. "High-Impact, Low-Frequency Event Risk to the North American Bulk Power System." June 2010. https://www.energy.gov/sites/prod/files/High-Impact%20Low-Frequency%20Event%20Risk%20to%20the%20North%20American%20Bulk%20Power%20System%20-%202010.pdf.

Phillips, Tony. "Solar Shield—Protecting the North American Power Grid." NASA, October 26, 2010. https://science.nasa.gov/science-news/science-at-nasa/2010/26oct_solarshield.

"Radiation from Solar Activity." Environmental Protection Agency. Accessed May 1, 2020. https://www.epa.gov/radtown/radiation-solar-activity.

Redd, Nola Taylor. "Space Weather: Sunspots, Solar Flares & Coronal Mass Ejections." Space.com, March 17, 2017. https://www.space.com/11506-space-weather-sunspots-solar-flares-coronal-mass-ejections.html.

"Space Weather Newspaper Archives." Solar Storms. Accessed May 1, 2020. http://www.solarstorms.org/SRefStorms.html.

MASS EXTINCTION

Anderson, Kate. "What's Normal: How Scientists Calculate Background Extinction Rate." Population Education, December 11, 2018. https://populationeducation.org/what-is-background-extinction-rate-how-is-it-calculated/.

"The Big Five Mass Extinctions." *Cosmos Magazine*. Accessed May 1, 2020. https://cosmosmagazine.com/palaeontology/big-five-extinctions.

Carrington, Damian. "Light Pollution Is Key 'Bringer of Insect Apocalypse.'" *The Guardian*, November 22, 2019. https://www.theguardian.com/environment/2019/nov/22/light-pollution-insect-apocalypse.

Carrington, Damian. "Plummeting Insect Numbers 'Threaten Collapse of Nature.'" *The Guardian*, February 10, 2013. https://www.theguardian.com/environment/2019/feb/10/plummeting-insect-numbers-threaten-collapse-of-nature.

Crockett, Lee. "A Coffin for Cod? The Downward Spiral of the Fish That Built New England." *National Geographic*, November 17, 2015. https://blog.nationalgeographic.org/2015/11/17/a-coffin-for-cod-the-downward-spiral-of-the-fish-That-built-new-england/.

"From Years of 'Miraculous Fishing' to Stock Collapse." IRD. Accessed May 1, 2020. http://www.suds-en-ligne.ird.fr/ecosys/ang_ecosys/intro2.htm.

Goldstone, Jack A. "Africa 2050: Demographic Truth and Consequences." Hoover Institution, January 14, 2019. https://www.hoover.org/research/africa-2050-demographic-truth-and-consequences.

Hanson, Amy Beth, and Matthew Brown. "Ex-Wildlife Chief: Trump Rule Could Kill Billions of Birds." Associated Press, March 31, 2020. https://apnews.com/a0565f7fc53c90ab5d78862a0135bd22.

Hiss, Tony. "Can the World Really Set Aside Half of the Planet for Wildlife?" *Smithsonian Magazine*, September 2014. https://www.smithsonianmag.com/science-nature/can-world-really-set-aside-half-planet-wildlife-180952379/.

"How Many Species on Earth? About 8.7 Million, New Estimate Says." Science Daily, August 24, 2011. https://www.sciencedaily.com/releases/2011/08/110823180459.htm.

International Whaling Commission. Accessed May 1, 2020. https://iwc.int/home.

Kourous, George. "Many of the World's Poorest People Depend on Fish." Food and Agriculture Organization of the United Nations. June 7, 2005. http://www.fao.org/newsroom/en/news/2005/102911/.

Levitt, Tom. "Overfishing Puts $42bn Tuna Industry at Risk of Collapse." *The Guardian*, May 2, 2016. https://www.theguardian.com/sustainable-business/2016/may/02/overfishing-42bn-tuna-industry-risk-collapse.

McKie, Robin. "Should We Give up Half of the Earth to Wildlife?" *The Guardian*, February 17, 2018. https://www.theguardian.com/environment/2018/feb/18/should-we-give-half-planet-earth-wildlife-nature-reserve.

Nature Beauty. "Amazing Automatic Lines, Catching and Processing Fish Right on Ship, Big Catch in the Sea." YouTube video, 13:49. May 5, 2019. https://www.youtube.com/watch?v=lBQCw4AYKaE.

"Pacific Bluefin Tuna." NOAA Fisheries. Accessed May 1, 2020. https://www.fisheries.noaa.gov/species/pacific-bluefin-tuna.

Plait, Phil. "Poisoned Planet." *Slate*, July 28, 2014. https://slate.com/technology/2014/07/the-great-oxygenation-event-the-earths-first-mass-extinction.html.

Platt, John R. "Lemurs in Crisis: 105 Species Now Threatened with Extinction." *Scientific American*, August 20, 2018. https://blogs.scientificamerican.com/extinction-countdown/lemurs-in-crisis-105-species-now-threatened-with-extinction/.

Ruz, Camila. "Amphibians Facing 'Terrifying' Rate of Extinction." *The Guardian*, November 16, 2011. https://www.theguardian.com/environment/2011/nov/16/amphibians-terrifying-extinction-threat.

Scheer, Roddy, and Doug Moss. "Are the World's Reptile Species in Trouble?" *Scientific American*, July 30, 2012. https://www.scientificamerican.com/article/reptiles-numbers-dwindling/.

"Science Study Predicts Collapse of All Seafood Fisheries by 2050." *Stanford Report*, November 2, 2006. https://news.stanford.edu/news/2006/november8/ocean-110806.html.

Wake, David B., and Vance T. Vredenburg. "Are We in the Midst of the Sixth Mass Extinction? A View from the World of Amphibians." *In the Light of Evolution, Volume II: Biodiversity and Extinction*, edited by John C. Avise, Stephen P. Hubbell, and Francisco J. Ayala, 27–44. National Academies Press: Washington, DC, 2008. https://www.ncbi.nlm.nih.gov/books/NBK214887/.

"Whaling Timeline." Greenpeace. Accessed May 1, 2020. http://www.greenpeace.org/usa/wp-content/uploads/legacy/Global/usa/planet3/PDFs/whalefacts-timeline.pdf.

World Food. "Here is The Big Catch—You Won't Believe That How Many Fishes—Amazing Fish Processing Machines." YouTube video, 10:30. November 14, 2008. https://www.youtube.com/watch?v=z8_T0csbx_I.

Zimmer, Carl. "Birds Are Vanishing from North America." *New York Times*, September 19, 2019. https://www.nytimes.com/2019/09/19/science/bird-populations-america-canada.html.

RAINFOREST COLLAPSE

Albert, Victoria. "Why the Amazon Rainforest Could Be at the Risk of 'Collapsing.'" *CBS News*, August 24, 2019. https://www.cbsnews.com/news/amazon-rainforest-fires-why-the-amazon-could-be-at-risk-of-collapsing-2019-08-24/.

"The Amazon Makes Its Own Wet Season." NASA. Accessed May 1, 2020. https://earthobservatory.nasa.gov/images/91161/the-amazon-makes-its-own-wet-season.

"Amazon Mining." World Wildlife Fund. Accessed May 1, 2020. https://wwf.panda.org/knowledge_hub/where_we_work/amazon/amazon_threats/other_threats/amazon_mining.

Barbosa, Vera. "Ever Wondered How Much Carbon Is Stored in a Tree?" CABI, June 30, 2011. https://cabiblog.typepad.com/hand_picked/2011/06/ever-wondered-how-much-carbon-is-stored-in-a-tree.html.

"Boreal Forest." Natural Resources Canada, February 25, 2020. https://www.nrcan.gc.ca/our-natural-resources/forests-forestry/sustainable-forest-management/boreal-forest/13071.

Butler, Rhett A. "Amazon Destruction." Mongabay, February 26, 2020. https://rainforests.mongabay.com/amazon/amazon_destruction.html.

Butler, Rhett A. "Calculating Deforestation Figures for the Amazon." Mongabay, April 24, 2018. https://rainforests.mongabay.com/amazon/deforestation_calculations.html.

Butler, Rhett A. "80% of Rainforests in Malaysia Borneo Logged." Mongabay, July 17, 2013. https://news.mongabay.com/2013/07/80-of-rainforests-in-malaysian-borneo-logged/.

Butler, Rhett A. "30% of Borneo's Rainforests Destroyed since 1973." Mongabay, July 16, 2014. https://news.mongabay.com/2014/07/30-of-borneos-rainforests-destroyed-since-1973/.

Chan, Melissa, and Heriberto Araújo. "China Wants Food. Brazil Pays the Price." *The Atlantic*, February 15, 2020. https://www.theatlantic.com/international/archive/2020/02/china-brazil-amazon-environment-pork/606601/.

"Climate Change Series Part 1—Rainforests Absorb, Store Large Quantities of Carbon Dioxide." Rainforest Trust, September 1, 2017. https://www.rainforesttrust.org/climate-change-series-part-1-rainforests-absorb-store-large-quantities-of-carbon-dioxide/.

"The Destruction of the Amazon, Explained." *The Week*, September 1, 2019. https://theweek.com/articles/861886/destruction-amazon-explained.

de Souza, Marcelo. "G7 Leaders Vow to Help Brazil Fight Fires, Repair Damage." *Associated Press*, August 25, 2019. https://apnews.com/452ac78bfea9484cb92e0cfb0b2234be.

FAPESP. "Amazon Deforestation Is Close to Tipping Point." Phys.org, March 20, 2018. https://phys.org/news/2018-03-amazon-deforestation.html.

Ferro, Shaunacy. "How Many Trees Are There in the Amazon?" *Mental Floss*, April 29, 2015. http://mentalfloss.com/article/63519/how-many-trees-are-there-amazon.

Francis, Nathan. "Two Weeks Ago, a Small Brazilian Newspaper Reported That Farmers Were Planning a 'Day of Fire' in the Amazon." *Inquisitr*, August 22, 2019. https://www.inquisitr.com/5594380/two-weeks-ago-a-small-brazilian-newspaper-reported-That-farmers-were-planning-a-day-of-fire-in-the-amazon/.

Freedman, Andrew. "Amazon Fires Could Accelerate Global Warming and Cause Lasting Harm to a Cradle of Biodiversity." *Washington Post*, August 22, 2019. https://www.washingtonpost.com/weather/2019/08/21/amazonian-rainforest-is-ablaze-turning-day-into-night-brazils-capital-city/.

Hurowitz, Glenn. "Here's What Deforestation Looks Like in 2019—and What We Can Do about It." Mighty Earth. Accessed May 1, 2020. http://www.mightyearth.org/heres-what-deforestation-looks-like-in-2019-and-what-we-can-do-about-it/.

Irfan, Umair. "Supertrees: Meet the Amazonian Giant That Helps the Rainforest Make Its Own Rain." *Vox*, December 23, 2019. https://www.vox.com/2019/12/12/20991590/amazon-rainforest-deforestation-climate-change-trees-rain-brazil-nut.

Kann, Drew. "The Amazon Is a Key Buffer Against Climate change. A New Study Warns Wildfires Could Decimate It." CNN, January 10, 2020. https://www.cnn.com/2020/01/10/world/amazon-rainforest-wildfires-climate-change-study/index.html

Levin, Kelly, and Katie Lebling. "CO2 Emissions Climb to an All-Time High (Again) in 2019." World Resources Institute, December 3, 2019. https://www.wri.org/blog/2019/12/co2-emissions-climb-all-time-high-again-2019-6-takeaways-latest-climate-data.

"Measuring the Daily Destruction of the World's Rainforests." *Scientific American*, November 19, 2009. https://www.scientificamerican.com/article/earth-talks-daily-destruction/.

"Mining's Big Environmental Footprint in the Amazon." Mining Technology, February 12, 2018. https://www.mining-technology.com/features/minings-big-environmental-footprint-amazon/.

"Modern Oil Palm Cultivation." Food and Agriculture Organization of the United Nations. Accessed May 1, 2020. http://www.fao.org/3/t0309e/T0309E01.htm.

Montaigne, Fen. "Will Deforestation and Warming Push the Amazon to a Tipping Point?" Yale Environment 360, September 4, 2019. https://e360.yale.edu/features/will-deforestation-and-warming-push-the-amazon-to-a-tipping-point.

Paddison, Laura. "From Algae to Yeast: The Quest to Find an Alternative to Palm Oil." *The Guardian*, September 29, 2017. https://www.theguardian.com/sustainable-business/2017/sep/29/algae-yeast-quest-to-find-alternative-to-palm-oil.

Pearce, Fred. "Rivers in the Sky: How Deforestation Is Affecting Global Water Cycles." Yale Environment 360, July 24, 2018. https://e360.yale.edu/features/how-deforestation-affecting-global-water-cycles-climate-change.

Phillips, Dom. "Illegal Mining in Amazon Rainforest Has Become an 'Epidemic.' *The Guardian*, December 10, 2018. https://www.theguardian.com/world/2018/dec/10/illegal-mining-in-brazils-rainforests-has-become-an-epidemic.

Piotrowski, Matt. "Nearing the Tipping Point." The Dialogue, May 2019. https://www.thedialogue.org/wp-content/uploads/2019/05/Nearing-the-Tipping-Point-for-website.pdf.

"Planting 1.2 Trillion Trees Could Cancel Out a Decade of CO2 Emissions, Scientists Find." Yale Environment 360, February 20, 2019. https://e360.yale.edu/digest/planting-1-2-trillion-trees-could-cancel-out-a-decade-of-co2-emissions-scientists-find.

Rasmussen, Carol. "New Study Shows the Amazon Makes Its Own Rainy Season." NASA, July 17, 2017. https://climate.nasa.gov/news/2608/new-study-shows-the-amazon-makes-its-own-rainy-season.

Rice, Doyle. "What would the Earth be like without the Amazon rainforest." *USA Today*, August 28, 2019. https://www.usatoday.com/story/news/nation/2019/08/28/amazon-rain-forest-what-would-earth-like-without-it/2130430001/.

Schlanger, Zoë. "The Global Demand for Palm Oil Is Driving Fires in Indonesia." *Quartz*, September 18, 2019. https://qz.com/1711172/the-global-demand-for-palm-oil-is-driving-the-fires-in-indonesia/.

Sergent, Jim, George Petras, and Elizabeth Lawrence. "6 Charts Show Why Thousands of Fires in the Amazon Rainforest Matter to the World." *USA Today*, January 10, 2020. https://www.usatoday.com/in-depth/news/2019/08/23/amazon-rainforest-six-charts-explain-why-fires-matter/2096257001/.

Shepherd, Marshall. "You Probably Know the Amazon Is Burning—Here's Why It Matters to You." *Forbes*, August 22, 2019. https://www.forbes.com/sites/marshallshepherd/2019/08/22/you-probably-know-the-amazon-is-burningheres-why-it-matters-to-you/#6732d4e4771a.

Shoumatoff, Alex. "Vanishing Borneo: Saving One of the World's Last Great Places." Yale Environment 360, May 18, 2017. https://e360.yale.edu/features/vanishing-borneo-saving-one-of-worlds-last-great-places-palm-oil.

"Where Are the Rainforests." Cal Tech. Accessed May 1, 2020. http://www.srl.caltech.edu/personnel/krubal/rainforest/Edit560s6/www/where.html.

"Which Is Worse for the Planet: Beef or Cars?" Eco Watch, July 13, 2016. https://www.ecowatch.com/which-is-worse-for-the-planet-beef-or-cars-1919932136.html.

Zimmer, Carl. "Never Mind That Boiling Kettle." *National Geographic*, January 12, 2004. https://www.nationalgeographic.com/science/phenomena/2004/01/12/never-mind-that-boiling-kettle/

HURRICANES AND TYPHOONS

Barnett, Cynthia. "Can We Engineer a Way to Stop a Hurricane?" *National Geographic*, October 13, 2017. https://www.nationalgeographic.com/news/2017/10/hurricane-geoengineering-climate-change-environment/.

"Coastal Foundations and Best Practices." Federal Emergency Management Agency. Accessed May 1, 2020. https://www.fema.gov/media-library-data/20130726-1707-25045-4311/chapter6.pdf.

Deziel, Chris. "The Average Wind Speed During a Thunderstorm." Sciencing, May 9, 2018. https://sciencing.com/average-wind-speed-during-thunderstorm-24075.html.

Dolce, Chris. "Top-10 Most Extreme Atlantic Hurricane Seasons in the Satellite Era." The Weather Channel, June 12, 2018. https://weather.com/storms/hurricane/news/top-10-most-extreme-hurricane-seasons.

"Family Flees Rising Water after Roof is Ripped Off Home." *CNN*. Accessed May 1, 2020. Video, 1:34. https://www.cnn.com/videos/weather/2019/09/01/bahamas-family-apartment-destroyed-nr-vpx.cnn.

Frazin, Rachel, and Tal Axelrod. "Death and Destruction: A Timeline of Hurricane Dorian." *The Hill*, September 7, 2019. https://thehill.com/homenews/news/460373-death-and-destruction-a-timeline-of-hurricane-dorian.

Georgiou, Aristos. "Scientists Are Planning to Stop Hurricanes in Their Tracks by Blocking Air Bubbles into the Sea." *Newsweek*, March 21, 2018. https://www.newsweek.com/scientists-are-planning-stop-hurricanes-their-tracks-blowing-air-bubbles-sea-855623.

"How Does a Hurricane Form?" SciJinks. Accessed May 1, 2020. https://scijinks.gov/hurricane/.

Jacobo, Julia. "Hurricane Andrew 25 Years Later: The Monster Storm That Devastated South Miami." *ABC News*, August 24, 2017. https://abcnews.go.com/US/hurricane-andrew-25-years-monster-storm-devastated-south/story?id=49389188.

Kaiser, Jocelyn. "A Sunshade for Planet Earth." *Science Magazine*, October 31, 2006. https://www.sciencemag.org/news/2006/10/sunshade-planet-earth.

Lai, K. K. Rebecca, Derek Watkins, Anjali Singhvi, Juliette Love, and Jugal K. Patel. "The Bahamas, Before and After Hurricane Dorian." *New York Times*, September 5, 2019. https://www.nytimes.com/interactive/2019/09/04/world/americas/bahamas-damage-hurricane-dorian.html.

Martinez, Gina. "'The Worst Natural Disaster I've Ever Seen.' U.S. Search and Rescue Team Describes Hurricane Dorian's Impact on the Bahamas." *Time*, September 13, 2019. https://time.com/5675918/florida-search-and-rescue-team-hurricane-dorian/.

Masters, Jeff. "Hurricane Dorian Was Worthy of a Category 6 Rating." *Scientific American*, October 3, 2019. https://blogs.scientificamerican.com/eye-of-the-storm/hurricane-dorian-was-worthy-of-a-category-6-rating/.

Pedersen, Joe Mario. "Hurricane Dorian Showed It's the Water Not the Wind as Storm's Deadliest Threat." *Orlando Sentinel*, September 19, 2019. https://www.orlandosentinel.com/weather/hurricane/os-ne-hurricane-dorian-flooding-storm-surge-water-killer-20190919-dho6moi4tjethif4ixcbnybasy-story.html.

Semple, Kirk. "Corpses Strewn, People Missing a Week After Dorian Hit the Bahamas." *New York Times*, September 8, 2019. https://www.nytimes.com/2019/09/08/world/americas/bahamas-dead-dorian.html?smid=nytcore-ios-share.

"Tropical Storms." Department of Atmospheric Sciences at University of Illinois at Urbana-Champaign. Accessed May 1, 2020. http://ww2010.atmos.uiuc.edu/(Gh)/guides/mtr/hurr/stages/ts.rxml.

OCEAN ACIDIFICATION

Carrington, Damian. "Ocean Acidification Can Cause Mass Extinctions, Fossils Reveal." *The Guardian*, October 21, 2019. https://www.theguardian.com/environment/2019/oct/21/ocean-acidification-can-cause-mass-extinctions-fossils-reveal.

Chu, Jennifer. "Ocean Acidification May Cause Dramatic Changes to Phytoplankton." *MIT News*, July 20, 2015. http://news.mit.edu/2015/ocean-acidification-phytoplankton-0720.

Clemets, Samantha. "The Faces and Functions of Algae on the Reef." Scripps Institution of Oceanography, February 27, 2015. https://scripps.ucsd.edu/labs/coralreefecology/the-faces-and-functions-of-algae-on-the-reef/.

Cummings, Colin. "How Acid Oceans Could Kill Krill (Op-Ed)." Live Science, July 18, 2013. https://www.livescience.com/38254-krill-collapse.html.

"FAQs about Ocean Acidification." Woods Hole Oceanographic Institution. Accessed May 1, 2020. https://www.whoi.edu/know-your-ocean/ocean-topics/ocean-chemistry/ocean-acidification/faqs-about-ocean-acidification/.

Leahy, Stephen. "This Gasoline Is Made of Carbon Sucked from the Air." *National Geographic*, June 7, 2018. https://www.nationalgeographic.com/news/2018/06/carbon-engineering-liquid-fuel-carbon-capture-neutral-science/.

"Mapped: The World's Coal Power Plants." Carbon Brief, March 26, 2020. https://www.carbonbrief.org/mapped-worlds-coal-power-plants.

Milman, Oliver, Christian Bennett, and Mike Bowers. "The Great Barrier Reef: An Obituary." *The Guardian*, March 26, 2014. https://www.theguardian.com/environment/ng-interactive/2014/mar/great-barrier-reef-obituary.

Nettle, Stu. "Ocean Acidification and the Rise of Jellyfish." Swellnet, December 11, 2015. https://www.swellnet.com/news/swellnet-dispatch/2015/12/11/ocean-acidification-and-rise-jellyfish.

"Ocean Acidification." Smithsonian. Accessed May 1, 2020. https://ocean.si.edu/ocean-life/invertebrates/ocean-acidification.

"Ocean Acidification Is Toxifying Phytoplankton." Planet Experts. Accessed May 1, 2020. http://www.planetexperts.com/ocean-acidification-toxifying-phytoplankton/.

"Ocean Acidification: Reducing CO2 Levels Is the Only Way to Minimise Risks." UNESCO. Accessed May 1, 2020. http://www.unesco.org/new/en/natural-sciences/ioc-oceans/infocus-oceans/features/ocean-acidification-2013/.

Qualman, Darrin. "It's Gonna Get Hot: Atmospheric Carbon Dioxide over the Past 800,000 Years." Darrin Qualman, February 14, 2017. https://www.darrinqualman.com/atmospheric-carbon-dioxide-co2/.

Scharf, Caleb A. "The Crazy Scale of Human Carbon Emission." *Scientific American*, April 26, 2017. https://blogs.scientificamerican.com/life-unbounded/the-crazy-scale-of-human-carbon-emission/.

Thomas, Emily. "U.S. Navy Has Found a Way to Turn Seawater into Fuel." Huffington Post, December 6, 2017. https://www.huffpost.com/entry/seawater-to-fuel-navy-vessels-_n_5113822.

Toh, Michelle. "Plankton Threatened by Ocean Acidification: Why That Matters." *Christian Science Monitor*, July 22, 2015. https://www.csmonitor.com/Science/Science-Notebook/2015/0722/Plankton-threatened-by-ocean-acidification-Why-That-matters.

Valentini, Kelli. "Acid Ocean: How Carbon Is Destroying the Shells of Marine Animals." NU Sci. Accessed May 1, 2020. https://nuscimag.com/acid-ocean-how-carbon-is-destroying-the-shells-of-marine-animals-11a09d624333.

GULF STREAM COLLAPSE

Bartelme, Tony. "A Powerful Current Just Miles from SC Is Changing. It Could Devastate the East Coast." *The Post and Courier*, September 5, 2018. https://www.postandcourier.com/news/special_reports/a-powerful-current-just-miles-from-sc-is-changing-it/article_7070df22-67fd-11e8-81ee-2fcab0fd4023.html.

Carrington, Damian. "Avoid Gulf Stream Disruption at All Costs, Scientists Warn." *The Guardian*, April 13, 2018. https://www.theguardian.com/environment/2018/apr/13/avoid-at-all-costs-gulf-streams-record-weakening-prompts-warnings-global-warming.

Cho, Renee. "Could Climate Change Shut Down the Gulf Stream." Earth Institute, June 6, 2017. https://blogs.ei.columbia.edu/2017/06/06/could-climate-change-shut-down-the-gulf-stream/.

Daley, Jason. "Ocean Current That Keeps Europe Warm Is Weakening." *Smithsonian Magazine*, April 13, 2018. https://www.smithsonianmag.com/smart-news/ocean-current-keeps-europe-warm-weakening-180968784/.

Fleshler, David. "The Gulf Stream Is Slowing Down. That Could Mean Rising Seas and a Hotter Florida." Phys.org, August 9, 2019. https://phys.org/news/2019-08-gulf-stream-seas-hotter-florida.html.

"Hurricane Movement." Windows to the Universe, March 31, 2009. https://www.windows2universe.org/earth/Atmosphere/hurricane/movement.html.

"Iceland's Climate Moves with the Gulf Stream Flow." Mountain Guides, January 23, 2015. https://www.mountainguides.is/blog/icelands-climate-moves-with-gulf-stream.

Lanchester, John. "How the Little Ice Age Changed History." *New Yorker*, March 25, 2019. https://www.newyorker.com/magazine/2019/04/01/how-the-little-ice-age-changed-history.

Miller, Elizabeth. "Deadly Currents—Why They Hit the Great Lakes." *WBFO*, August 16, 2016. https://news.wbfo.org/post/deadly-currents-why-they-hit-great-lakes.

Mules, Ineke. "Gulf Stream System at Weakest Point in 1,600 Years." DW, November 4, 2018. https://www.dw.com/en/gulf-stream-system-at-weakest-point-in-1600-years/a-43348456.

Niiler, Eric. "Can Tiny Glass Beads Keep Arctic Ice From Melting? Maaaybe." *Wired*, October 18, 2019. https://www.wired.com/story/geoengineering-tiny-glass-beads-prevent-arctic-ice-from-melting/.

"Paradoxically, One Outcome of Global Warming Could Be a Dramatic Cooling of Britain and Northern Europe." Hyper History. Accessed May 1, 2020. https://www.hyperhistory.com/online_n2/connections_n2/climate_n2/gulfstream.html.

"Photo Gallery: October Tidal Flooding in Hampton Roads and NE N.C." *WTKR*, October 11, 2019. https://wtkr.com/2019/10/11/photo-gallery-october-tidal-flooding-in-hampton-roads/.

"Streamcode—Planktonic Diversity of the Gulf Stream." Smithsonian Ocean. Accessed May 1, 2020. https://ocean.si.edu/planet-ocean/tides-currents/streamcode-planktonic-diversity-gulf-stream.

"Surface Ocean Currents." NOAA. Accessed May 1, 2020. https://oceanservice.noaa.gov/education/kits/currents/05currents1.html.

Wilcox, Jennifer. "A New Way to Remove CO2 from the Atmosphere." Filmed April 2018 at TED2018, Vancouver, Canada. Video, 14:08. https://www.ted.com/talks/jennifer_wilcox_a_new_way_to_remove_co2_from_the_atmosphere?language=en.

Zimmerman, Kim Ann. "What Is the Gulf Stream?" Live Science, January 15, 2013. https://www.livescience.com/26273-gulf-stream.html.

ROBOT TAKEOVER

"AI Beats Professionals in Six-Player Poker." *Science Daily*, July 11, 2019. https://www.sciencedaily.com/releases/2019/07/190711141343.htm.

"Albert HUBO." Hanson Robotics. Accessed May 1, 2020. https://www.hansonrobotics.com/albert-hubo/.

Barkho, Gabriela. "Walmart Confirms Use of AI-Powered Cameras to Detect Stealing." *Observer*, June 27, 2019. https://observer.com/2019/06/walmart-ai-cameras-detect-stealing/.

"Brain Neurons & Synapses." The Human Memory, September 27, 2019. https://human-memory.net/brain-neurons-synapses/.

Lewis, Dynai. "Where Do We Come From? A Primer on Early Human Evolution." *Cosmos Magazine*, June 9, 2016. https://cosmosmagazine.com/palaeontology/where-did-we-come-from-a-primer-on-early-human-evolution.

Lewis, Tanya. "Human Brain: Facts, Functions & Anatomy." Live Science, September 28, 2018. https://www.livescience.com/29365-human-brain.html.

"Met Police to Deploy Facial Recognition Cameras." *BBC*, January 30, 2020. https://www.bbc.com/news/uk-51237665.

Molina, Brett. "Ford Wants to Use Walking Robots to Help Self-Driving Cars Deliver Packages." *USA Today*, May 22, 2019. https://www.usatoday.com/story/money/cars/2019/05/22/ford-walking-robot-deliver-packages/3765162002/.

"The Neuron." Brain Facts, April 1, 2012. http://www.brainfacts.org/brain-anatomy-and-function/anatomy/2012/the-neuron.

"Overview of Neuron Structure and Function." Khan Academy. Accessed May 1, 2020. https://www.khanacademy.org/science/biology/human-biology/neuron-nervous-system/a/overview-of-neuron-structure-and-function.

Reber, Paul. "What Is the Memory Capacity of the Human Brain?" *Scientific American*, May 1, 2010. https://www.scientificamerican.com/article/what-is-the-memory-capacity/.

Tzezana, Roey. "Singularity: Explain It to Me Like I'm 5-Years-Old." Futurism, March 3, 2017. https://futurism.com/singularity-explain-it-to-me-like-im-5-years-old.

"Watson." IBM. Accessed May 1, 2020. https://www.ibm.com/watson.

ALIEN INVASION

Baggaley, Kate. "How Many Stars Are There in the Whole Galaxy?" *Popular Science*, October 24, 2018. https://www.popsci.com/how-many-stars-are-there-in-whole-galaxy/.

"Exoplanet and Candidate Statistics." NASA Exoplanet Archive. Accessed May 1, 2020. https://exoplanetarchive.ipac.caltech.edu/docs/counts_detail.html.

"Habitable Exoplanets Catalog." Planetary Habitability Laboratory, January 16, 2020. http://phl.upr.edu/projects/habitable-exoplanets-catalog.

Hendricks, Scotty. "Dark Forest Theory: A Terrifying Explanation of Why We Haven't Heard from Aliens Yet." Big Think, June 14, 2018. https://bigthink.com/scotty-hendricks/the-dark-forest-theory-a-terrifying-explanation-of-why-we-havent-heard-from-aliens-yet.

"How Do We Know How Many Galaxies Are in Our Universe?" Physics.org. Accessed May 1, 2020. http://www.physics.org/facts/sand-galaxies.asp.

ImageworksVFX. "Edge of Tomorrow—Creating the Mimics Shot Build." YouTube video, 2:51. June 17, 2015. https://www.youtube.com/watch?v=iiKeTPL6HPk.

"Kepler's Science Results." NASA. Accessed May 1, 2020. https://exoplanets.nasa.gov/resources/2189/keplers-science-results/.

Morelle, Rebecca. "Dolphins 'Call Each Other by Name.'" *BBC*, July 23, 2013. https://www.bbc.com/news/science-environment-23410137.

Parker, Laura. "Rare Video Shows Elephants 'Mourning' Matriarch's Death." *National Geographic*, August 31, 2016. https://www.nationalgeographic.com/news/2016/08/elephants-mourning-video-animal-grief/.

Steele, Bill, and Linda B. Glaser. "Google Celebrates Arecibo Message to Extraterrestrials." Cornell University, November 15, 2018. https://astro.cornell.edu/news/google-celebrates-arecibo-message-extraterrestrials.

Wall, Mike. "Number of Habitable Exoplanets Found by NASA's Kepler May Not Be So High After All." Space.com, October 28, 2018. https://www.space.com/42275-habitable-exoplanets-kepler-discoveries-revised-by-gaia.html.

RELATIVISTIC KILL VEHICLE

Cain, Fraser. "How Does the Sun Produce Energy?" Phys.org, December 14, 2015. https://phys.org/news/2015-12-sun-energy.html.

Dockrill, Peter. "NASA's Working on a Nano-Starship That Travels at 1/5 the Speed of Light." Science Alert, December 9, 2016. https://www.sciencealert.com/nasa-s-working-on-a-nano-starship-That-travels-at-1-5-the-speed-of-light.

Felder, Adam. "Battering the Batter." *The Atlantic*, May 5, 2015. https://www.theatlantic.com/entertainment/archive/2015/05/no-more-battering-the-batter/391991/.

Gibbs, Philip. "The Relativistic Rocket." University of California, Riverside, Department of Mathematics. Accessed May 1, 2020. http://math.ucr.edu/home/baez/physics/Relativity/SR/Rocket/rocket.html.

Hadhazy, Adam. "How Fast Could Humans Travel Safely through Space?" *BBC*, August 10, 2015. https://www.bbc.com/future/article/20150809-how-fast-could-humans-travel-safely-through-space.

Jimiticus. "Interstellar Travel: Approaching Light Speed." YouTube video, 12:00. December 21, 2016. https://www.youtube.com/watch?v=c4z6RZXv5p8.

Johnson-Groh, Mara. "Three Ways to Travel at (Nearly) the Speed of Light." NASA, May 29, 2019. https://www.nasa.gov/feature/goddard/2019/three-ways-to-travel-at-nearly-the-speed-of-light.

Powell, Corey S. "These New Technologies Could Make Interstellar Travel Real." *Discover Magazine*, March 12, 2019. https://www.discovermagazine.com/the-sciences/these-new-technologies-could-make-interstellar-travel-real.

"Relativistic Baseball." What If?. Accessed May 1, 2020. https://what-if.xkcd.com/1/.

Rice, Doyle. "1,000-mph Winds, Shock Waves Deadliest Effects of Asteroid Strike." *USA Today*, April 19, 2017. https://www.usatoday.com/story/tech/sciencefair/2017/04/19/asteroid-strike-deadliest-hazards/100652436/.

"The Sun's Energy." UTIA. Accessed May 1, 2020. https://ag.tennessee.edu/solar/Pages/What%20Is%20Solar%20Energy/Sun's%20Energy.aspx

Timmer, John. "Just How Dangerous Is It to Travel at 20% the Speed of Light?" *Ars Technica*, August 23, 2016. https://arstechnica.com/science/2016/08/could-breakthrough-starshots-ships-survive-the-trip/.

SciNews. "Breakthrough Starshot–Nanocraft to Alpha Centauri." YouTube video, 4:28. April 12, 2016. https://www.youtube.com/watch?v=RoCm6vZDDiQ.

"What If an Asteroid Hit the Earth?" Glenn Learning Technology Project. Accessed May 1, 2020. https://www.grc.nasa.gov/www/k-12/Numbers/Math/Mathematical_Thinking/asteroid_hit.htm.

GRAY GOO

Ball, Philip. "Make Your Own World with Programmable Matter." Institute of Electrical and Electronics Engineers Spectrum, May 27, 2014. https://spectrum.ieee.org/robotics/robotics-hardware/make-your-own-world-with-programmable-matter.

Edwards, Barry. "The Genetics of Viruses and Bacteria." Slide Player. Accessed May 1, 2020. https://slideplayer.com/slide/9519439/.

Flemming, Nic. "Which Life Form Dominates Earth?" *BBC*, February 10, 2015. http://www.bbc.com/earth/story/20150211-whats-the-most-dominant-life-form.

Gasser, Charles. "Primary Classes of Molecules in *E. coli*." University of California, Davis. Accessed May 1, 2020. https://www.mcb.ucdavis.edu/courses/bis102/ecoli.html.

Heddings, Anthony. "What Do '7nm' and '10nm' Mean for CPUs, and Why Do They Matter?" How-To Geek, January 22, 2019. https://www.howtogeek.com/394267/what-do-7nm-and-10nm-mean-and-why-do-they-matter.

Herring, Angela. "The Rising Red Tide with Climate Change." Phys.org, May 16, 2013. https://phys.org/news/2013-05-red-tide-climate.html.

"How Big Is an *E. coli* Cell and What Is Its Mass?" Bionumbers. Accessed May 1, 2020. http://book.bionumbers.org/how-big-is-an-e-coli-cell-and-what-is-its-mass.

Jones, Scotten. "TSMC and Samsung 5nm Comparison." SemiWiki, May 3, 2019. https://semiwiki.com/semiconductor-manufacturers/samsung-foundry/8157-tsmc-and-samsung-5nm-comparison/.

"Sperm." *Molecular Biology of the Cell, 4th Edition* by Alberts, B., A. Johnson, J. Lewis, Martin Raff, Keith Roberts, and Peter Walter, 432. Garland Science: New York, 2002. https://www.ncbi.nlm.nih.gov/books/NBK26914.

Todar, Kenneth. "Nutrition and Growth of Bacteria." Todar's Online Textbook of Bacteriology. Accessed May 1, 2020. http://textbookofbacteriology.net/nutgro.html.

Welsh, Jennifer. "'Dumped' Pythons Put Squeeze on Everglades Wildlife." Live Science, January 30, 2012. https://www.livescience.com/18192-invasive-pythons-everglades-wildlife.html.

"What Is Cancer." National Cancer Institute, February 9, 2015. https://www.cancer.gov/about-cancer/understanding/what-is-cancer.

Zimmer, Carl. "How the Brown Rat Conquered New York City (and Every Other One, Too)." *New York Times*, October 27, 2016. https://www.nytimes.com/2016/10/28/science/brown-rat-new-york-city.html.

PICTURE CREDITS

INDEX

ABOUT THE AUTHOR

Marshall Brain is best known as the founder of HowStuffWorks.com, an award-winning website eventually purchased by Discovery Channel.

Marshall is a well-known public speaker with the ability to deliver complex material in an easygoing, understandable way. He has appeared on radio and TV programs nationwide. He has been featured on outlets such as CNN, *Good Morning America*, *Modern Marvels*, and *The Oprah Winfrey Show*. He has also hosted the primetime show *Factory Floor with Marshall Brain* on the National Geographic Channel.

In addition to creating HowStuffWorks, Marshall is the author of more than a dozen books as well as a number of widely known web publications. His books include *The Engineering Book* (Sterling), *How "God" Works* (Sterling), *How Stuff Works* (Wiley), *More How Stuff Works* (Wiley), *What If?* (Wiley), *How Much Does the Earth Weigh?* (Wiley), *The Teenager's Guide to the Real World* (BYG Publishing), and *Motif Programming* (Digital Press). His web projects include Manna—Two Views of Humanity's Future, Imagining Elon Musk's Million Person Mars Colony, The Day You Discard Your Body, The Second Intelligent Species, and Robotic Nation.

Marshall earned a Bachelor of Science in electrical engineering from Rensselaer Polytechnic Institute and a Master of Science in computer science from North Carolina State University. He lives in Cary, North Carolina, with his wife and four kids and is the director of the Engineering Entrepreneurs Program at North Carolina State University.